# ENERGY SERFDOM
## TO
# ENERGY FREEDOM

**UNDERSTANDING PETROLEUM SOURCES
AND ISSUES IN THE 21$^{ST}$ CENTURY**

## Donald R. Statter, Jr.

Energy Serfdom to Energy Freedom
© 2015 by Donald R. Statter, Jr.
All rights reserved.

Cover Design: Hyliaan Graphics
Cover Photo Attribution: © www.123rf.com/profile_3dmentat'
Interior Formatting: The Author's Mentor,
          www.LittleRoniPublishers.com

ISBN-13:978-1508616382
ISBN-10:1508616388
Also available in eBook

PUBLISHED IN THE UNITED STATES OF AMERICA

# Table of Contents

Chapter 1: Introduction ........................................................ 1

Chapter 2: Petroleum Basics ................................................. 5

Chapter 3: Natural Gas ....................................................... 29

Chapter 4: Other Petroleum Sources .................................... 52

Chapter 5: Coal .................................................................. 57

Chapter 6: The Case for Fischer-Tropsch ............................ 61

Chapter 7: Review of Petroleum Sources ............................. 67

Chapter 8: Methane Hydrate ............................................... 71

Chapter 9: United States Petroleum History ....................... 75

Chapter 10: Supply-Demand Dynamics .............................. 103

Chapter 11: Petroleum Refining ......................................... 107

Chapter 12: Environmental Mishaps ................................... 115

Chapter 13: Renewable Energy ........................................... 122

Chapter 14: Facing Challenges ........................................... 130

Chapter 15: Conclusion ..................................................... 139


Information Sources .......................................................... 144

About the Author .............................................................. 151

# 1

## *Introduction*

---

It doesn't matter what the issue is, if it deals with energy or the environment, some combination of lies, politics, and misinformation drive the debate. Petroleum sources and related issues in the 21st Century are no exception. The world isn't running out of petroleum, its running out of truth.

What would you do if you ran a business and agents remote from your control said things and did things that drove up the value of your product? We'll revisit the issue drivers related to this question in greater detail later on in the book, but for now let's just leave you the reader with the understanding that the petroleum industry will not challenge misrepresentations that drive up the value of its products.

There's an information gap between the petroleum producers and petroleum consumers. The producers use this gap to take advantage of consumers. Many of you that are interested in the title of this book want to know, "is there a plentiful supply of petroleum today that will last long into the future?" How about price dynamics, "should we expect the price of petroleum to keep rising long into the future?"

Petroleum prices fluctuate on a daily basis. The mid-July 2014 value for West Texas Intermediate, the benchmark petroleum as traded within the

*The world isn't running out of petroleum, its running out of truth.*

United States, sold for approximately $100 per barrel. By the end of year 2014 the price of petroleum dropped to under $60 per barrel.

Oil prices throughout the book are provided in United States Dollar (USD) currency value of the day. A second value in parenthesis is the re-calculated value in 2010 USD.

This book is intended to answer the questions referenced above, but will start with the basics so that the reader will understand the origin of petroleum; the volume of petroleum in the United States in reference to the rest of the world; key factors that drive the value of petroleum and the potential for synthetic petroleum alternatives in the future. This book will provide the information you need so that you can become a better informed consumer. If enough of the public reads this book it could lead toward the opportunity for a collective response that serves as a virtual national energy policy. This nation dearly needs a national policy that makes energy independence a priority.

This nation suffers from many issues including a perceived energy crisis. Our national leaders and the petroleum industry gain power and wealth through this misinformation. A better informed public will have the knowledge to understand that we are not running out of petroleum any time soon, and yes we are generally paying too much for it. Petroleum is an energy source that remains vital to the nation and the broader world, too vital to allow it to be manipulated by lies, politics and misinformation.

This book will also cover environmental issues of concern such as the emerging shale gas industry, fracking, and new petroleum sources that will be viable later this century. While we're discussing the environment, did you know it was the petroleum industry that saved the whales from extinction . . . how can that be?

In the mid-19th Century the primary source for fuel oil, lubricants, and wax was extracted from the blubber and acoustic cavities of whales. The petroleum age, as we currently know it, began with the first successful oil well drilled by Edwin Drake, near Titusville, Pennsylvania, in 1859. It's beyond the scope of this book to describe the transition from whale oil to petroleum in the 19th Century, as this is a book about the 21st Century. But the reader needs to understand that the decline of the whaling industry began with that singular event.

The shift from whale oil to petroleum didn't occur because of overwhelming public outcry, new government regulations, or a legal process decided by the courts. Whales were saved from extinction because petroleum made them obsolete as an oil source. It was an economic decision, as whale oil was too expensive, unreliable as a source, and the supply could not meet the demands of a growing industrial base. There are many lessons to be learned from that element of history.

Two centuries later we need to ask these same questions about the petroleum supply. Is petroleum obsolete? Can petroleum continue to meet the energy demands of a growing world industrial base? Is petroleum a reliable source of energy, specifically in the United States; can we produce enough petroleum or synthetic equivalent to meet our needs without dependence on foreign sources? What about the cost? Economies stabilize when they can control the factors that impact production costs such as petroleum based energy. Can petroleum be an affordable energy source long into the future? This book will shed some light on these issues.

# 2

## *Petroleum Basics*

Petroleum is a naturally occurring hydrocarbon. That means it's primarily composed of carbon and hydrogen, but may contain minor amounts of other elements such as sulfur, nitrogen, or oxygen. Generally a liquid, this book will also include petroleum discussions related to natural gas, and hydrocarbons created by synthetic processes.

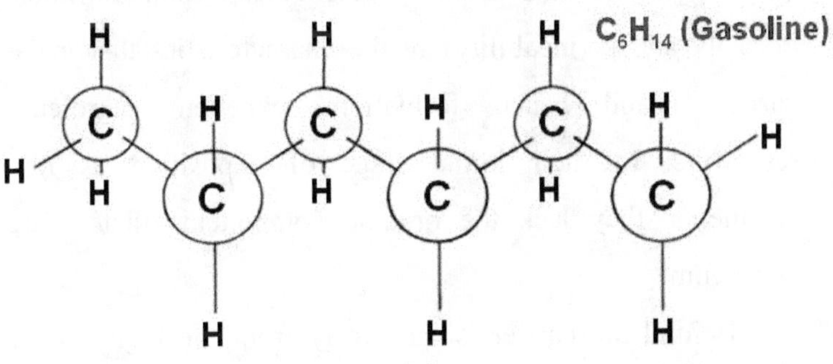

Figure 1. Simple Hexane chain, one of the gasoline series of chemicals

5

## Petroleum Origin

Sand bodies of coastal origin were the source for 61% of the world's petroleum before shale gas extraction techniques were developed in the late 1990's. To understand what this means think of the barrier islands that are common along the Atlantic Coast from New England, to the southern tip of Florida, and continues to trend north and west along the coast of the Gulf of Mexico, through to Texas. The United States coastline from the Atlantic Ocean, through to the Gulf of Mexico represents a continuous series of barrier islands over 3,000 miles long.

The barrier island sequence is characterized by coarse, clean sand and gravel overlaying fine grained marine mud. The sand and gravel deposits are the beach and sand dunes that represents that portion of the barrier island that lies toward the high energy ocean environment. These sediments are clean when deposited, and they generally have good porosity, or void space; and permeability, or interconnections that allow for the transmission of fluids. Porosity and permeability are the characteristics that make barrier island beach formations excellent petroleum reservoirs. In their initial stage of deposition as raw sediments, they lack the organic components that create petroleum.

Behind the barrier island, away from the high energy environment characterized by the ocean, is the back-bay. The sediment deposited in the low energy back-bay environment is generally fine grained mud. The mud is dark and foul smelling because a major component of the sediment includes organic materials contributed from decayed plants and animals. Much of these organic components are microscopic algae, other phytoplankton, and zooplankton.

To better understand the difference between sand and marine mud you have to visualize the sediments in microscopic detail. Figure 2 is a generalized illustration of microscopic thin sections of marine mud and sand. Marine mud is composed of thin clay plates, in contrast to sand which is generally composed of quartz spheroids. The marine mud appears as a stack of thin bricks because the illustration is a cross-section. The plates have a well ordered orientation because the flat side of each plate has an opposing positive and negative electric charge, so that it acts like a weak, but tiny magnate.

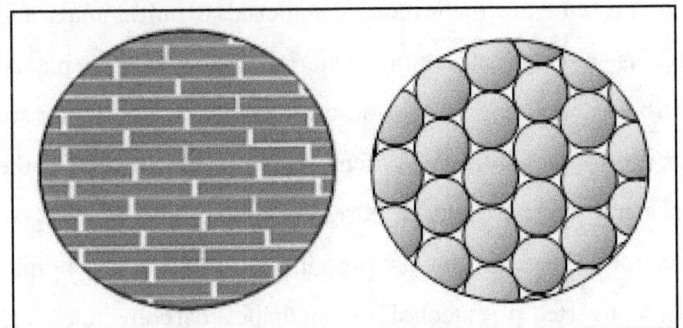

**Figure 2. Microscopic schematic of marine mud (left) and sand (right)**

When deposited in sea water, the plates that compose marine mud are tightly compacted because the salt in the water is an electrolyte that compels the microscopic plates to stick to one another due to electronic attraction. This tight compaction reduces the porosity and permeability of marine mud, in contrast to beach sand and gravel deposits that generally have good porosity and permeability. Marine mud ultimately becomes a sedimentary rock type termed shale when it becomes lithified due to the heat and pressure of deep burial within the earth.

Within the barrier island model, marine organisms that will ultimately become petroleum, such as algae and zooplankton, are buried in the low energy back-bay reducing environment. That means these materials decay without oxygen. The organic compounds that ultimately become petroleum are the proteins, lignins, carbohydrates, and lipids that comprise marine life.

Proteins are the structural materials from animals and lignins are the structural materials from higher plants. Carbohydrates are energy sources from sugar and lipids are energy sources from fat. When organic sediment such as the back-bay marine mud is buried and deprived of oxygen, anaerobic bacteria convert proteins to amino acids; lignins are converted to alcohol or methane; carbohydrates are converted to sugars; and lipids are converted to fatty acids. This anaerobic process is similar to digestion in higher

8

animals and represents the first step in petroleum genesis, the formation of kerogen.

Petroleum genesis occurs simultaneously with subsidence, compaction, and lithification of the source bed sediments. Kerogen will become a major component of the marine mud as the sediments subside deeper into the earth. This kerogen is cooked and migrates from the fine grained back-bay sediment where it was deposited, into the coarse grained beach sediment where it can be contained and extracted by well drilling in the future.

Primary migration occurs because the marine mud source bed loses its capacity to contain the petroleum it has produced. In the traditional petroleum paradigm, marine mud, or shale, starts as a source bed, and converts over time to a confining layer that forms a petroleum trap.

The most basic form of petroleum trap is the anticline. The anticline is a structural ridge, or arch. The axis of the anticline is the highest portion of the structure. As petroleum forms and migrates out of fine grained source rock, it moves up, typically into anticlines until it is trapped by an impermeable confining layer. In the traditional approach to petroleum exploration, shale forms the confining layers, and sands form the petroleum reservoirs.

Within the petroleum reservoir, gas lies above oil, and oil lies above connate water. Connate water is the original sea water that was deposited with the sediments. This water contained the organic compounds that ultimately became petroleum, which is lighter than water, so it migrates to the highest point in the reservoir structure.

Temperature and depth relationships are critical to understanding petroleum genesis. Diagenesis is the low temperature reactions that take place below 50°C (122°F), which equates to depths to 2,500 feet. These are low temperature reactions that begin to remove oxygen, sulfur, and nitrogen from the organic compounds that were

*Thirty-six states in the United States have known petroleum reserves.*

a product of the initial anaerobic decay processes. The primary products of diagenesis are kerogen, heavy crude, and biogenic gas.

Catagenesis is the term for intense petroleum genesis which occurs at temperatures ranging from 50°C (122°F) to 200°C (392°F), or a depth range from 2,500 to 15,000 feet. The optimum depth for petroleum genesis is at 5,000 to 10,000 feet. In the catagenic temperature zone of the earth the conversion of kerogen to complex chains of carbon and hydrogen is completed. Natural gas, chains of 1 to 5 carbons, is pervasive at all depths within a sedimentary basin. Heavy crude, which contains significant amounts of tar or wax, made of chains of carbon generally greater than 30 in number, is found from near the earth's surface to 15,000 feet.

Tar and wax are complex organic chemicals. They are rarely found together in the same oil reservoir. Tar is created from source beds where the kerogen was created from microscopic marine organisms. Wax is created from source beds where the kerogen was created from terrestrial vegetation. Light crude is chains of 6 to 30 carbons, is generally found from 2,500 to 15,000 feet deep.

Metamorphosis is the terminal stage of petroleum genesis which occurs at temperatures above 200°C (392°F) where petroleum is converted to coal or graphite. There is no liquid petroleum to be found below 15,000 feet.

## Petroleum Production

Thirty-six states in the United States have known petroleum reserves. The eastern Gulf of Mexico and off-shore Atlantic Coast are two regions with strong potential for petroleum that currently have no production. Even without petroleum contributions from these regions the United States was the number three oil producer in the world in 2013.

For the year 2013, the United States produced 7.54-million Barrels of oil per Day (B/D); behind Saudi Arabia at 9.4-million B/D; and Russia at 10.40-million B/D. By the time this book goes to print the United States will be the number one oil producer in the world, but complete statistics for 2014 do not yet exist. The United States has the potential to produce significantly more oil with a positive impact on supply, but the typical United States oil well is too small to have a positive impact on price.

| ## | COUNTRY | 2013 OIL PRODUCTION (B/D) | NUMBER OF WELLS | AVERAGE WELL (B/D) |
|---|---|---|---|---|
| 1 | Russia | 10,403,900 | 124,497 | 84 |
| 2 | Saudi Arabia | 9,379,300 | 2,895 | 3,240 |
| 3 | United States | 7,534,800 | 460,221 | 16 |
| 4 | China | 4,211,300 | 71,552 | 59 |
| 5 | Canada | 3,327,600 | 82,661 | 40 |
| 6 | Iraq | 3,225,700 | 1,526 | 2,114 |
| 7 | Kuwait | 2,562,000 | 1,286 | 1,992 |
| 8 | Iran | 2,554,800 | 2,074 | 1,232 |
| 9 | Abu Dhabi | 2,543,600 | 1,200 | 2,120 |
| 10 | Mexico | 2,530,200 | 8,315 | 304 |
| 11 | Venezuala | 2,465,300 | 14,651 | 168 |
| 12 | Brazil | 2,092,200 | 11,792 | 177 |
| 13 | Nigeria | 1,910,400 | 2,110 | 905 |
| 14 | Angola | 1,752,100 | 1,335 | 1,312 |
| 15 | Kazakhstan | 1,618,900 | 1,256 | 1,289 |
| 16 | Norway | 1,488,800 | 814 | 1,829 |
| 17 | Libya | 1,233,500 | 2,060 | 599 |
| 18 | Algeria | 1,143,800 | 2,014 | 568 |
| 19 | Columbia | 1,018,600 | 7,159 | 142 |
| 20 | RoW | 12,281,300 | 93,831 | 131 |
| | World | 75,278,100 | 893,249 | 84 |

**Figure 3. Oil well performance by country**

According to figure 3, Saudi Arabia stands alone as the leader in oil well performance, with the average well producing 3,200 B/D. The nations of Abu Dhabi and Iraq are a distant second, with oil production of 2,100 B/D. The world average for petroleum well performance is 84 B/D.

The United States is the leader in the number of oil wells with over 460,000 wells that produce a modest average of 16 B/D. The United States has a significant petroleum infrastructure where smaller wells can be justified as an investment because the equipment to complete production wells travels a shorter distance with lower risk than similar projects in other parts of the world. In remote regions of the world the oil fields must be significantly larger in order to justify the cost and risk of petroleum exploration.

## Ghawar Oil Field, Saudi Arabia

It takes 200 United States oil wells to match the production of a single Saudi Arabian oil well. To understand the magnitude of oil production from Saudi Arabia we need to understand the Ghawar Oil Field which is considered to be the largest in the world.

The Ghawar Oil Field is 175 miles long and 20 miles wide and accounts for 60% of the Saudi Arabian oil production. It produces 5-million B/D or 6.5% of the world oil production in 2009. As of the year 2010, the Ghawar Oil Field had produced 65-billion barrels of oil, yet it is believed to have reserves of over 100 billion barrels of oil remaining.

In addition to oil, the Ghawar Oil Field produces 330,000 barrels of oil equivalent per day (BOE/D) of natural gas. Natural gas is generally measured in cubic feet ($Ft^3$) or cubic meters ($M^3$), but this book will convert natural gas production figures to the Barrel of Oil Equivalent (BOE) unit to reduce the ambiguity between oil and natural gas production figures. Later in the book there will be a description of how the BOE unit is derived.

The first Ghawar Oil Field discovery well was completed in 1947. Oil production began in 1951. Production increases were slow and steady throughout the decades of the 1950's and 1960's so that oil production stood at 2.6-million B/D by 1970.

The Organization of Petroleum Exporting Countries (OPEC), a cartel headquartered in Vienna, Austria, was formed in 1960. That means its members through collective processes attempt to control a market through supply manipulations in order to maximize profits. Although many OPEC nations are Muslim religion or Arab culture, many are not. The common link for membership within OPEC is an economy based on oil exports.

In the first decade of their existence OPEC had a negligible impact on petroleum prices. World oil supplies were high, costs were low, and the world economy grew accordingly. The Ghawar Oil Field production trend steadily increased throughout the decade of the 1960's as the world oil price trend hit historic lows by 1972.

The first collective response that had a major impact on world oil prices occurred in 1973 as a result of the 1st Oil Embargo related to the Yom Kippur War. Arab nations such as Saudi Arabia did not cut total oil production as a result of the war, but temporary cuts, in combination with nation specific export restrictions caused a dramatic shift in oil prices. The major action that caused world oil cost increases was to stop the export of oil to the allies of Israel, such as the United States. Oil prices increased because the United States, the world's major oil consumer market and importer of petroleum lost control of its supply from the world's major export nations. Ghawar Oil Field production peaked at 5-million B/D in 1974 when the price of oil peaked above $9 ($40) per barrel. The Saudis used the 1st Oil Embargo as an opportunity to profit from Arab politics. After thirteen years of operation, OPEC was now relevant and in control of the world oil supply and price structure by use of the 1973 Oil Embargo.

Throughout the mid-1970's the price of oil remained high, but steadily dropped until the Iranian Revolution, which began in 1978 with reduced oil shipments from the Persian Gulf. Saudi oil production from the Ghawar oil field wasn't purposely cut, but the supply depended on transport from tanker ships heading through the Persian Gulf and the Strait of Hormuz, and the Iranian Revolution threatened the viability of this choke point. The supply disruption caused temporary production decreases and related price increases.

Soon after the Shah of Iran was deposed and a new Islamic republic was declared, Iraq invaded Iran. It's beyond the scope of this book to describe details of the war or the motives behind it, but it needs to be understood that a result of the war was periodic supply disruptions for oil shipments from the Persian Gulf. The net effect of the war was oil price increases that resulted in peak production motivations when safe transport opportunities were possible. Ghawar Oil Field production peaked at 5.3-million B/D when the price peaked above $37 ($106) per barrel in 1980.

After 1981, the rest of the world caught up with the Persian Gulf oil supply problem. Ten years of oil price instability were followed by 20 years of relative oil price stability and world economic growth. One of the critical factors that enable prosperity in the industrialized world is affordable energy. Economies bloom when even the lowest class members of society can afford to travel to and from work, and that requires low cost fuel.

So what happened to cause the Oil Glut of the 1980's? Many factors contributed to the glut such as new oil supplies from the non-OPEC North Sea nations of Norway and the

United Kingdom. In the United States, oil from the north slope of Alaska was reaching the world market via the Trans-Alaskan Oil Pipeline that opened in 1977. Another great source of United States oil production was the offshore platforms within the Gulf of Mexico.

Consumer behaviors also had a significant impact on supply-demand dynamics. By the late 1970's American drivers

> *It takes 200 United States oil wells to match the production of a single Saudi Arabian oil well.*

began to drive smaller, more fuel efficient cars. Smaller cars in combination with a recession that began in 1980 resulted in reduced driving habits that had positive effects for consumers on the world oil price.

But the greatest single impact to increase the petroleum supply and lower the price was the end to the oil price control structure that was imposed on domestic oil production in 1973. As a result of the price control structure, oil from new domestic production wells sold at the fixed price of $32 ($77) per barrel, and oil sold from pre-existing wells sold for $40 ($97) per barrel.

One of the first acts of President Ronald Reagan in 1981 was to end the two tiered price control structure. Domestic oil exploration and production boomed for a short period of time with the expectation that the regulatory controls were keeping the oil price down. But history proves the reverse was true. Domestic oil companies that drilled new wells anticipating price increases as a result of deregulation quickly learned that the price controls imposed by the federal government were artificially high, not artificially low.

New petroleum sources outside OPEC control, combined with a recession that began in 1980, and adjustments to driver behaviors had a positive impact on oil prices for consumers. After 1981 the world price for oil dropped from historic highs of over $37 ($106) per barrel in 1980 to under $15 ($31) per barrel by 1986. Ghawar Oil Field production dropped accordingly, the Saudis were not going to contribute to world over production which dropped to 1.1-million B/D by 1985, the lowest production level recorded since 1967.

After 1985 the Ghawar Oil Field followed a gradual trend in production increases that peaked at 5.2-million B/D in 1997. This production peak coincided with a world oil price of less than $12 ($18) per barrel. After 1997, production decreased over the next two years, so the drop in Ghawar Oil Field production to 4.6-million B/D coincided with the late 1990's oil price increase to $17 ($23) per barrel in 1999. No other nation has more effect on oil supply and price dynamics than Saudi Arabia, and the Ghawar Oil Field is the premier oil field that gives the Saudis this power.

18

## Petroleum Cost Factors

Remember the question from the Introduction? What would you do if you ran a business and agents remote from your control said things, and did things that drove up the value of your product? So what are some of the misrepresentations that drive up the value of petroleum?

Peak Oil is a theoretical concept postulated by a Geophysicist by the name of M. King Hubbert, in 1956. According to his theory the world has a finite amount of oil and would suffer from a decline in affordable access to petroleum resources within the next half-century. While it's true that the amount of petroleum is finite, his theory did not account for the immense volumes of petroleum that exist in the world, the great variety of regions from which it can be extracted, new drilling and extraction technologies, and new sources of petroleum or synthetic equivalents.

With very few exceptions, every region of the world that had oil production in 1950's, when the theory was postulated, is still producing oil today. We know from our discussion of Saudi Arabian oil in the last section of the book that they have led the world in oil production over the past 40 years, and will likely maintain that claim for the next 40 years, unless politics or economic decisions shift their priorities.

It's true that the United States domestic oil production appeared to follow a Peak Oil trend when production reached 10.2-million B/D in the 1970's, then declined throughout the 1980's. This oil production decline was an effect of economic decisions and politics, not a true lack of petroleum. Foreign oil from non-OPEC sources was cheaper to produce because the domestic wells from the United States were smaller and less profitable than the foreign oil that was imported. By the 1980's the major source of oil for the United States was off-shore drilling platforms in the Gulf of Mexico and the north slope of Alaska where oil production was high enough to compete with foreign sources of oil.

In many cases politics will drive a region toward a premature production peak. The Santa Barbara region of California has the potential for more oil production, but environmental concerns drives policy away from increasing petroleum production. The north slope of Alaska has been a reliable high-volume oil production region since the 1970's. In contrast to California, the locals in the state, including the Inupiat Tribe that lives along the Arctic Coast, support petroleum exploration and production increases. Alaska appears to be following a Hubbert Peak Oil production curve, not because more petroleum doesn't exist in the region, but because national environmental politics from outside the region is resisting an increase in drilling activity against the will of the local residents.

Another fallacy impacting perceived supply-demand dynamics is the concept that North America in general and the United States specifically cannot produce enough petroleum, and has no choice but to import petroleum from foreign nations to meet market demands. This just isn't true. The concept that the United States can't produce enough petroleum to meet its own domestic market demand is another misrepresentation driven by politics.

Although they likely know otherwise, petroleum industry professionals in the United States and foreign suppliers such as the OPEC cartel will not disagree with this claim because the lie supports their goal to maintain high values for their product. To really understand the issue as it relates to the United States petroleum supply, you need to understand national energy policies that were established in the post World War II period over 60 years ago.

A critical element of winning World War II was the great volume of oil that was produced domestically from oil fields from states such as Texas, Louisiana, Oklahoma, Pennsylvania and California. Although these oil fields served their mission well, by the end of the war they appeared to be nearly spent and new sources were needed to feed the post-war economic expansion. Oil had yet to be discovered in Alaska and the technology to allow deep off-shore drilling had yet to be developed.

The United States looked to the Middle-East for oil where the supply appeared to be cheap, plentiful, and stable well into the future. This strategy was established before the creation of OPEC and the advent of radical Islam. It was a reasonable and appropriate strategy for the time, but times have changed. But before we move on, Texas oil is not spent as it was perceived after World War II; 2012 production figures confirm that Texas is the leading producer of oil within the

> *One myth that harms the value and viability of petroleum as an energy source is the idea that carbon dioxide is a pollutant that causes global warming.*

United States with a current production figure of 2.0-million B/D.

As it stands today (2013 production figures) the United States has a petroleum consumption rate of 18.5-million B/D, of which 7.4-million B/D, or 40%, needs to be imported. While it is true the United States currently stands as number twelve on the list of nations in reference to known petroleum reserves, the United States has the potential to find new discoveries within our own boundaries to reduce our need for imports.

According to government estimates released in 2006 by the federal agency formerly known as the Mineral Management Service, now called the Bureau of Ocean Energy Management, undiscovered technically recoverable oil and gas resources from off-shore United States sources could be increased by a dramatic 86-billion barrels of oil plus 89-billion BOE of natural gas. But the Gulf of Mexico east of 87° longitude and all of the Atlantic coast continental shelf are off-limits to petroleum exploration due to political pressure. In addition there is an estimated 49 billion barrels that could be developed from new on-shore discoveries such as the Williston Basin of North Dakota and the north slope of Alaska.

*Man-made global warming is a politically motivated lie.*

Another myth that harms the value and viability of petroleum as an energy source is the idea that carbon dioxide is a pollutant that causes global warming. This cannot be true. Carbon dioxide is a part of the respiration equation used by all animal life forms since the beginning of creation. A similar equation is used with the burning of organic fuels such as petroleum. When oxygen combines with an organic fuel such as sugar, wood, or a fossil fuel it creates water plus carbon dioxide. The carbon dioxide and water enters the atmosphere where it becomes a critical resource nutrient for plants. Plants reverse the respiration process used by animals by the use of sunlight to create oxygen plus sugars from carbon dioxide and other nutrients. To consider carbon dioxide a pollutant is to believe that our Creator failed when He set in motion the basic processes that control life in the Universe.

Now we've been told by the environmental establishment not to burn wood or other carbon fuels because the process releases carbon dioxide into the atmosphere. Although this may be true, it also needs to be understood that a rotting tree releases the same amount of carbon dioxide to the atmosphere as a burning tree, it just happens much slower. A real study of the effects of extra carbon dioxide in the atmosphere is more likely to confirm that plants grow faster; but that concept would not satisfy the goals of environmental politics.

Real science suggests that global warming, if or when it occurs, is a product of sun spot activity, not carbon dioxide. The environmental establishment wants to hide the solar global warming relationship because the sun is remote from human control. John Coleman, the founder of the Weather Channel, is one of many meteorology professionals that believe the whole carbon dioxide and earth temperature issue is a hoax. For more information you can read his report *"The Amazing Story Behind the Global Warming Scam"* that describes how a single scientist by the name of Roger Revelle started the issue in 1958 in an effort to justify more funding for his government sponsored research efforts.

Although the Revelle research efforts spawned new environmental programs including the United Nation's Intergovernmental Panel on Climate Change, he did not remain an advocate for his initial claims. In 1988 he wrote two cautionary letters to members of Congress stating, *"My own personal belief is that we should wait another 10 or 20 years to really be convinced that the greenhouse effect is going to be important for human beings, in both positive and negative ways."* He added, *". . . we should be careful not to arouse too much alarm until the rate and amount of warming becomes clearer."*

Have you ever stopped to realize what part of our atmosphere is composed of

*The petroleum industry will not challenge misrepresentations that drive up the value of its products.*

carbon dioxide? When studies began in 1958, carbon dioxide was measured at 315 parts per million and has since increased to 385 parts per million by 2008. To put this concept into perspective there are about 600 molecules of oxygen in the atmosphere for every one of carbon dioxide, and that ratio has remained fairly constant throughout the age of this debate. If you could talk to an oak tree it would likely tell you that the earth suffers from too much oxygen and craves more carbon dioxide.

Man-made global warming is a politically motivated lie used to justify new taxes, more government control, and extreme regulation over humanity. Just like the fear of barbarian invasion was used as a means of making serfs out of free citizens after the collapse of Rome. In the present day, the environmental establishment wants to make energy serfs out of common citizens by scaring us into believing the world is cycling into a dangerous warming trend caused by man. To paraphrase Benjamin Franklin, "A man who would trade his *energy* freedom for a *false* security *based on a lie*, deserves neither".

The petroleum industry doesn't believe we're running out of oil or it would be working harder to lead consumers toward alternative energy systems. Other lies and misrepresentations are not harming the industry but contributing to the value of its product because the enemies of petroleum want it to be more expensive in order to reduce its consumption. The petroleum industry will not challenge misrepresentations that drive up the value of its products.

Before we close this section of the book we need to establish real numbers that lead toward an understanding of the production costs for petroleum. Each region of the world has its own unique factors that drive the cost of production. To understand the cost of petroleum we will study the high risk, high reward offshore platforms that operate in the Gulf of Mexico.

There are over 4,500 petroleum production platforms operating in the federally controlled off-shore region of United States Gulf of Mexico. One of these platforms is called the Na Kika Project, jointly owned by Royal Dutch Shell and British Petroleum, 144 miles southeast of the city of New Orleans.

The Na Kika Project is considered to be the largest off-shore petroleum platform project of its kind in the world. From one platform location in 6,000 feet of water, directional drilling techniques are used to connect the platform with six separate oil fields, up to 15 miles away.

According to the Na Kika Project website, $1.4-billion will be spent to produce 300-million barrels of oil over the planned duration of the project. A calculation of the money spent verse the volume of oil produced leads to the conclusion that the final cost of production comes to $4.70 per barrel. This is not a bad profit when you consider oil generally trades at over $100 per barrel.

Before we run off and think we're getting ripped-off because we're all paying too much for oil, we need to understand the cost of time and risk associated with off-shore petroleum production. Off-shore oil production is typically a 30-year venture. The first ten years is spent with planning prime off-shore locations through geophysical science and engineering studies, to include permit procurements from the federal government, and extraordinary actions related to environmental politics. There's also likely to be some banking involved; it doesn't matter how big a single company or set of joint venture partners are, few organizations are large enough to spend over a $1-billion without borrowing the funds.

27

If all goes according to plan, a successful offshore platform project is ready to place the oil platform and begin drilling within ten years of conception. The second ten years will be spent drilling a multitude of wells to remote locations by use of directional drilling techniques and replacing the drill string with production pipelines that bring the petroleum to surface locations where it can be delivered to tanker ships or distribution pipelines. In general it should take about 20 years for an offshore oil platform to break even in reference to the cost of investment verse the value of the oil produced. If the project is successful, the last ten years are spent making profits, so by year 30 production figures that were claimed at the beginning of the project should match the volume of petroleum that was produced.

Off-shore platform drilling is a big risk, big reward process. The companies that perform these types of projects use the revenue from mature platforms to pay the bank notes on the young platforms that aren't producing yet. When a company has to wait 30 years for a pay-off, the consumer can't expect the oil company to use its profits to cut prices; but the off-shore petroleum industry is potentially lucrative if and when they're able to put more than $1-billion at risk and can wait 30 years for a pay-off.

# 3

## *Natural Gas*

Throughout this book natural gas volumes will be measured in a unit called the Barrel of Oil Equivalent (BOE). The first step in understanding how the BOE is derived is to understand the British Thermal Unit (BTU). The basic definition of a BTU is the amount of heat energy required to raise one pound of water, 1°F. A BOE of natural gas has the same BTU value as one 42 gallon barrel of crude oil. According to the United States Geologic Survey (USGS), one barrel of crude oil has the heat energy value of 6,000,000 BTU, or 6,000 cubic feet ($Ft^3$) of natural gas, which equates to 170 cubic meters ($M^3$) of natural gas.

Natural gas is a series of five organic chemicals: methane ($CH_4$), ethane ($C_2H_6$), propane ($C_3H_8$), butane ($C_4H_{10}$), and pentane ($C_5H_{12}$). These organic chemicals are

29

called Alkanes. That means they are saturated hydrocarbons composed of carbon and hydrogen as there is only a hydrogen or a carbon at every bond position within their respective molecular structures. Commonly used as a fuel, natural gas can also be used as a feedstock for building larger liquid or solid molecules of higher complexity through refining processes.

Figure 4 illustrates a simple anticline in map view and cross-section view. There are hundreds, if not thousands of these simple anticline structures throughout the Texas and Louisiana Gulf Coast region. They are generally a mile or so in diameter and might not have a surface expression that appears visible by analysis of terrain or topography. They are typically created from minor salt intrusions, called diapirs that create subtle folds, vice major salt intrusion events that are indicated by bold surface features called salt domes.

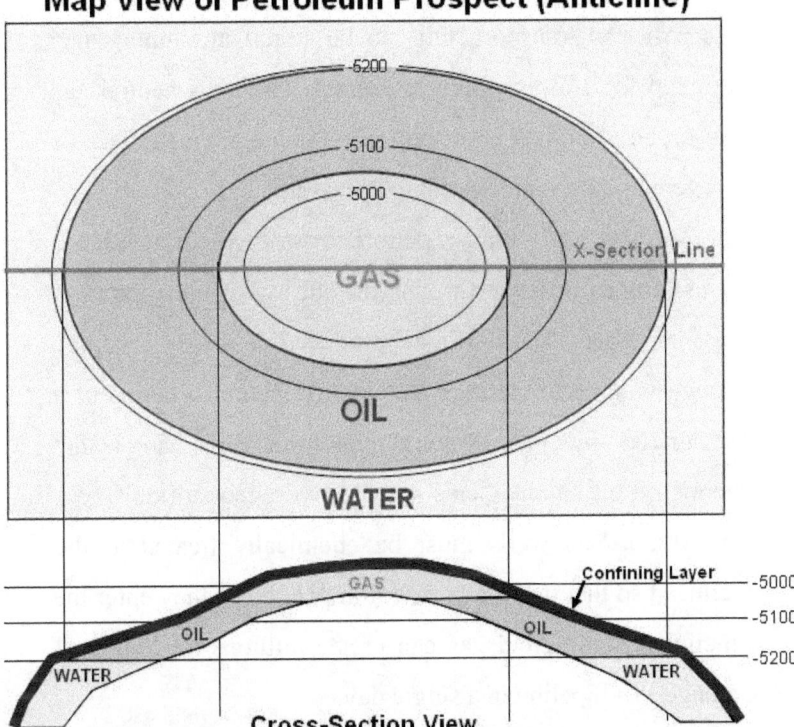

Figure 4. Map and Cross-Section of simple anticline with natural gas

The primary purpose of figure 4 is to illustrate how gas lies over oil, and oil lies over water within the petroleum reservoir. The water in this graphic is termed connate water. Connate water is the same water that was originally trapped within the original sediment deposits. Remember that the major source of petroleum constituents originated from microscopic organisms that were deposited in sea water.

In the traditional petroleum production paradigm, oil is preferred before natural gas because it has a higher value and it is easier to transport. Oil can be stored in simple tanks designed for 1 atmosphere of pressure, whereas natural gas has to be stored or transported as a compressed gas or cryogenic liquid under very high pressure.

The gas cap within a petroleum reservoir is preserved for as long as possible to push oil out by natural formation pressure. Natural gas will be produced if there is a large enough volume in the region to justify the construction of a pressurized pipeline. Natural gas can have an acidic chemistry due to chemicals such as hydrogen sulfide ($H_2S$). These sour gas wells must be chemically treated at the wellhead so that they have a neutral pH before they enter the pipeline system. Acid gas can create millions of dollars of damage to a pipeline in a single day.

## Natural Gas Production

World natural gas production stands at 70.2-million BOE per day (BOE/D). The United States is the number one producer of natural gas in the world with 2012 production of 11-million BOE/D, or 15.6% of the world supply. This volume of natural gas is possible because the United States has vast volumes of natural gas plus an extensive pipeline infrastructure to bring it to market.

Russia is the number two natural gas producer in the world with production of 10.8-million BOE/D, and Iran is a distant third place natural gas producer with 2.6-million BOE/D.

## Recent Natural Gas and Oil Price History

Figure 5 illustrates the price of natural gas in comparison to that of oil from 1997 through early 2014. Except for winter demand spikes in 2001 and 2003, the BOE cost of natural gas remained similar to the price of a barrel of oil through year 2005. A third natural gas price spike occurred in late 2005 as a result of Hurricane Katrina, when Gulf Coast platforms had to shut down production operations as a mitigation strategy to prevent related storm damage. After 2008 the price of natural gas in reference to the price of oil began to diverge. Since 2010 the BOE price of natural gas generally trades at 30% of the value of a barrel of oil.

The dominant price spike in figure 5, called the "Oil Price Bubble" occurred in 2008 when the price of oil rose to over $140 per barrel and natural gas followed at $80 per BOE by July of that year. The price of petroleum rapidly dropped in the second half of the year 2008.

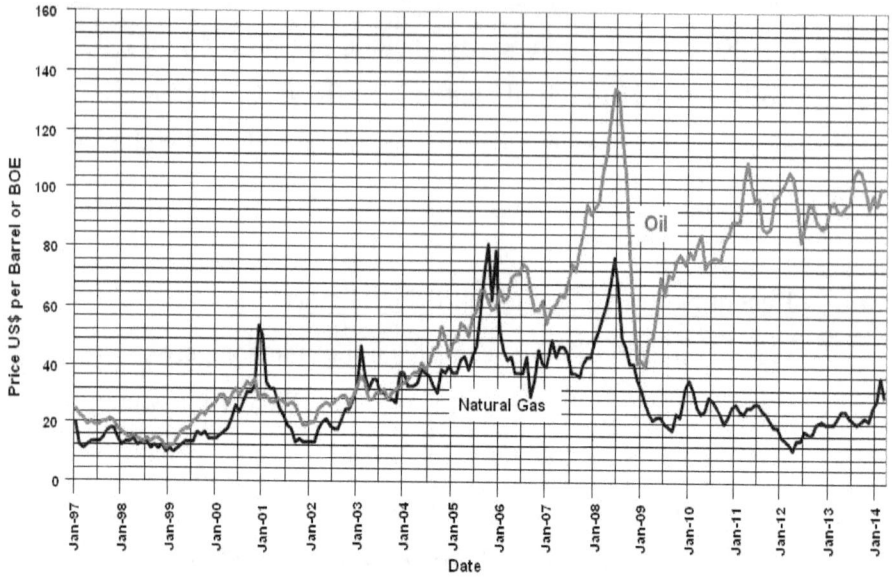

**Figure 5. Oil and Natural Gas price trend, 1997 through early 2014**

What was the cause of the price spike? The first cause to be considered is the Peak Oil concept that was described earlier in the book. Was the world running out of oil and the world supply could not keep pace with the world demand? The short answer is no, the Peak Oil concept that the world is running out of oil just doesn't fit the trend. In the same year that the price spiked to historic highs, the price dropped to historic lows. The world had essentially the same number of oil consumers at the beginning of the year as existed at the end of the year. The only difference is that the price spike modified consumer behavior. People stopped driving or drove more fuel efficient cars when they had the choice. This happens every time there's a dramatic oil price spike.

What about mid-east war supply disruptions? Although the United States continued to have military operations in Iraq and Afghanistan in 2008, there was nothing unique that occurred in 2008 that hasn't occurred before or since with different effects. There were no unique war induced supply disruptions to cause the 2008 oil price spike.

A third factor that needs to be considered is futures trading. In 2002 there were 4.5 barrels of oil traded in paper for every real barrel of oil produced in the world. Although this ratio seems very unhealthy, the paper trading trend got significantly worse after 2002. By mid-2008 there were 14.5 shares of oil traded on paper for every real barrel of oil produced in the world.

The oil price trend after 2002 is a close match to the paper trading trend increase over the same period of time. Paper trading increased by 300% and the price of oil increased by the same amount over the same time period accordingly. The price of oil dropped like a rock after July 2008, when the United States Congress scheduled hearings for August of that year to uncover the cause and consider solutions to the oil speculation scandal. Fear of Congress had a positive effect, by the end of the year the price of oil dropped to $40 per barrel. Unfortunately the world price for oil increased to just over $100 per barrel by mid-2014 since early 2009 and there remains just over 15 paper shares of oil traded for every real barrel of oil produced in the world.

The price of oil was over valued by mid-2014 because it was over traded. We know from history that over trading a commodity can lead to wild price fluctuations. Commodities that are over traded are either over valued, or under valued, but rarely priced at or near their real value. Think of the 1929 Stock Market Crash that was caused by margin trading.

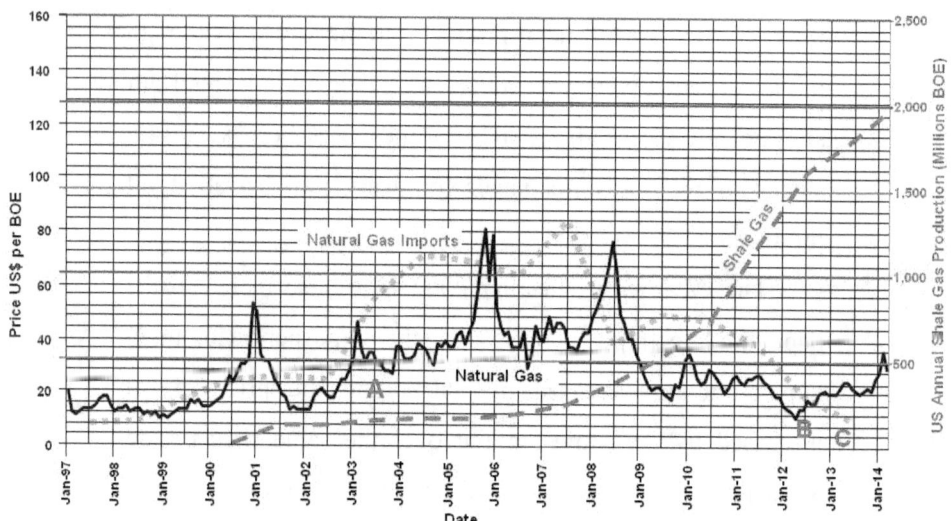

**Figure 6. Natural Gas price, Shale Gas and Import Trends**

The problem with paper trading in oil is that the traders never have to take delivery of the product they've purchased in paper. These traders can't take delivery of the products as they don't have million barrel oil tanks in their back yards to store the oil they've purchased. Maybe the solution to the oil trading problem is to insist that oil traders can only be companies or individuals that have storage capacity to receive the product they're trading. In other words, an oil trader has to have a million barrels in tank capacity if he wants the right to purchase a million barrels of oil on paper.

Now let's explore natural gas price dynamics. As explained earlier in the book, the price of oil and natural gas began to diverge after 2008. What happened in 2008 to bring the price of natural gas down to where it trades for 30% of the value of oil?

> *Supply-demand dynamics allow the United States to be the low cost natural gas producer in the world.*

Figure 6 is similar in content to figure 5 except that the oil price trend is removed and two other trend lines are superimposed. The first trend line that needs to be referred to is the dashed line that represents United States shale gas production. According to the graphic, shale gas production began in year 2000 and has had a steady increase to 2-billion BOE per year by mid-2014. By the time United States shale gas production reached 500-million BOE per year in 2009 the natural gas price dropped to under $30 per BOE, where it has remained for the past five years.

The United States shale gas supply is plentiful, cheap and continues to increase. Supply-demand dynamics allow the United to be the low cost natural gas producer in the world. To better understand this concept we need to examine the dotted line superimposed on figure 6. This dotted line represents Liquid Natural Gas (LNG) imports to the United States. According to the graphic, LNG imports peaked when the world natural gas price was relatively low in 2004 and 2007, before shale gas production was significant after 2009. LNG imports to the United States dropped significantly in 2008 when natural gas prices peaked along with the price of oil during the price bubble event. Since mid-2009 the trend for LNG imports to the United States has dropped significantly to 1997 levels, below 200-million BOE per year.

There are eleven LNG processing plants in the United States and Puerto Rico. These are off-shore or near shore facilities where LNG ships connect to pipeline terminals in order to import or export natural gas. One of these facilities that have been in the news throughout much of 2014 is the Cove Point LNG Plant in Lusby, Maryland. What happens at this plant is of special interest to the local residents that live near the plant, but should be of general interest to the rest of the United States. The Cove Point LNG Plant began imports of natural gas in 2003 after several decades of dormancy as indicated by the letter 'A' labeled on figure 6. After nine years of LNG imports, the Cove Point facility ended natural gas imports in 2012, as indicated by the letter 'B' in the graphic, when prices were at a historic low. This event took

38

place because the price of the natural gas that was being imported through the LNG facility was more expensive than the natural gas being produced in the market it was intended to supply.

Cove Point LNG operations are motivated by economic factors. The LNG terminal cannot justify gas imports that are more expensive than the market they are attempting to supply and remain in business. Early in 2013 the Cove Point LNG Plant filed for an export license with the federal government as indicated by the letter 'C' on the graphic. Similar to the magnitude of the off-shore petroleum platforms discussed earlier in the book, upgrades to allow LNG exports is a $3.5-billion project that has mixed support from the local community. If or when the project is complete, the Cove Point LNG Plant will be the first export facility of its kind in the lower 48 states. No other LNG export facilities exist, planned, or pending along the Atlantic coast or Gulf of Mexico. Lusby, on the Chesapeake Bay in southern Maryland, is an excellent geographic location for serving as an export facility for the transport of emerging Marcellus Shale gas fields that are being developed throughout West Virginia and Pennsylvania. If the Cove Point facility fails to complete the LNG export project due to political pressure it will likely close.

## Fracking

Fracking is short for the term hydraulic fracturing and is an engineering process to induce permeability of an otherwise tight petroleum production zone by breaking or cracking the rock formation to stimulate petroleum flow into the well casing. Fracking is performed in order to improve petroleum production in wells that would otherwise be dry holes.

Originally employed as a mitigation for local conditions to stimulate the flow of petroleum that were generally not predicted prior to locating a well, the use of the technique with horizontal drilling is now a planned or pre-meditated technique used to extract petroleum. When fracking is used with horizontal drilling it allows for the extraction of petroleum at extended distances remote from the well location. Horizontal drilling is used to increase the length of interaction contact within the producing horizon by drilling along the formation, not through the formation. In the oil and gas industry, fracking has become common practice, with beneficial and detrimental effects.

In the 1860's, oil well operators found that using force to fracture the rock around a shallow, hard rock oil well increased the productivity of the well. Fracturing the rock released trapped petroleum products for extraction by enhancing the permeability, or fluid transmission capability, of the rock formation. Initially, this was accomplished by lowering dynamite or nitroglycerin charges into the well and exploding them. On April 25, 1865, Col. Edward A. L. Roberts received the first of many patents for an "exploding torpedo" designed for that purpose. Col. Roberts' technique was adopted by oil producers in Pennsylvania, New York, Kentucky, and West Virginia, using nitroglycerin. The same technique was later applied in water and natural gas wells. In the 1930's, oil producers pumped hydrochloric acid into fractured wells to reduce the likelihood of the cracks that were formed from closing.

In 1964, the U.S. Bureau of Mines began research on the use of explosive fracturing to enhance the formation of rock fractures in order to extract petroleum from shale. A research paper, published by the Bartlesville Energy Research Center, showed that nitroglycerin could be used to cause extensive fracturing to expose more of the petroleum-containing shale to enable faster petroleum extraction.

Although still used in mining and quarrying, explosive fracturing has limited application in oil and gas well operations, because its effects are more localized than hydraulic fracturing. Also, cost and safety are important considerations.

Hydraulic fracturing, the technique most widely used today, was introduced to the petroleum industry in 1947 as a result of research conducted by Floyd Farris of the Stranolind Oil and Gas Company. In the initial experiment, oil drillers injected 1,000 gallons of gelled gasoline and sand into a gas producing limestone well in southwest Kansas. Although the experiment was less than successful, because the well's productivity didn't significantly increase, Stranolind felt the idea had merit. In 1948, J.B. Clark of Stranolind published a paper describing the process, and an exclusive patent for it was issued in 1949. The Halliburton Oil Well Cementing Company negotiated with Stranolind and acquired the exclusive license to use the process. In March, 1949, Halliburton employed the process commercially for the first time, in two wells located in Oklahoma and Texas. Since then, fracking has been successfully used to stimulate a million or more oil and gas wells in various geologic formations.

In 1952, oil producers in the Soviet Union began the use of propping agents in fracking. Propping agents are solid materials, such as sand or man-made beads, added to the fracking fluid in order to keep the rock fractures open. Propping agents can be added during or after the fracturing process. The use of propping agents has been widely adopted by the fracking industry.

In 1968, Pan American Petroleum performed high-volume fracking in Oklahoma. Although the definition varies, high-volume fracking often involves pumping more than 100,000 pounds of propping agents into a well.

In the 1980's, oil well operators in Texas began

42

completing many wells by drilling horizontally into the Austin Chalk; they also used high-volume fracking. This horizontal well approach optimizes access to much more of the petroleum resources contained in the rock formation. Consider that a rock formation that is 50 feet in the vertical could be 100's or 1000's of feet thick in the horizontal dimension by use of horizontal drilling techniques.

As we learned in the second chapter of the book, shale, or rock made from fine grained marine mud deposits, generally have a low volume of voids, termed porosity, and low fluid transmission properties, termed permeability. In the process of petroleum genesis the organic materials that are buried in the sediments that become shale, migrate to sandstone deposits as they convert to petroleum. Until recently, shale has been ignored for petroleum exploration and production, because it has low porosity and even lower permeability. But with the introduction of high-volume fracking in shale, the rock is fractured by force and huge volumes of petroleum are potentially unlocked. George P. Mitchell applied this approach in shale formations for the first time in 1997. Mitchell has been called "the father of fracking" as a result. Since then, high-volume fracking has been adopted on a commercial scale to extract petroleum products from shale formations in the United States, Canada, and China. Several other nations are planning to employ the technique in unconventional oil and gas production.

Hydraulic fracturing is accomplished by the insertion of a tube and packing into a petroleum well casing. The tube is used to pump a pressurized mixture of gelled water and other substances into the rock strata, causing it to crack. The tube includes packing, or a "packer", in order to prevent backflow of the fracking fluid and supports the compression necessary to fracture the rock formation.

Fracking fluid is typically 99% or more water by volume. However, it contains a variety of chemical compounds which are injected into the rock formation. Used in quantities up to 8,000,000 gallons, the fluid includes: acids, salts, lubricants, disinfectants, gelling agents, radioactive tracer isotopes, solvents and propping agents.

After fracking is complete, some of the mixture is lost into the fractured rock created in the process, leaving the propping agents in place to keep the fractures open. Most of the fluid flows back to the surface as pressure dissipates. This process is called "flowback", which is followed by additional water which was naturally present in the rock formation, termed "connate water", which we learned about earlier in the book. Both the flowback and the connate water contain the fracking fluids listed above, in addition to a small amount of naturally-occurring radioactive material. When the process is complete nitrogen is injected into the well to force the recovery of as much of the fracking fluid as possible.

The current shale gas boom uses high-volume fracking technique that is similar to the traditional process, except that it employs a horizontal geometry that increases the area of impact. In high-volume fracturing, a single

> *Fracking is producing gas in regions that used to be ignored.*

well may be fractured in several stages, repeating the pressurization and backflow process multiple times to induce dramatic increases in petroleum production.

The petroleum industry is currently working under the paradigm that:

1. Fracking is required in some regions of the United States or there would be no gas or oil production;

2. Fracking is producing gas in regions that used to be ignored, or dead to traditional petroleum production;

3. Fracking is allowing the US to make great steps toward energy independence;

4. Fracking is reducing the risk of dry holes, and;

5. Fracking is increasing the volume of petroleum production.

In some regions of the United States such as the offshore Gulf Coast, gas and oil continue to be produced and can be extracted using conventional wells without fracking. Fracking appears to be the default approach for other regions of the country such as the Bakken Shale of North Dakota and the Marcellus Shale of West Virginia and Pennsylvania. Conventional wells are the alternative to fracking and should be used wherever it makes sense to do so. Fracking should be employed when drilling into rock formations composed of impermeable material such as shale, but should not be used in areas where the formations are permeable and petroleum is plentiful by traditional means, such as sandstones and some porous limestone formations.

High-volume fracking has practical applications in specific geologic formations, but should not be the standard approach to extracting petroleum from all locations. The capability of fracking with horizontal drilling to extract petroleum at remote distances from the mapped well location provides the possibility of mineral rights avoidance, impact to protected areas, and damage to the public trust.

| CRITERIA | ~1980 | ~2014 |
|---|---|---|
| PRIMARY TARGET | Oil | Natural Gas |
| SECONDARY TARGET | Natural Gas | Oil |
| STRUCTURE | Anticline | Doesn't Matter |
| LITHOLOGY | Sandstone or Pourous Limestone | Shale |
| SHALE? | Confining Layer | Petroleum Source |
| FRACKING? | Mitigation for Localized Conditions | Intentional Process |
| SUCCESS RATE | ~30% | ~100% |

**Figure 7. Comparison of fracking motivations, 1980 and 2014**

The motivation shift that has occurred within the American petroleum industry toward shale gas techniques since 1980 is described in figure 7. As the graphic depicts, oil was the primary target 34 years ago, gas was secondary, because at that time the United States did not have the pipeline infrastructure needed to bring natural gas to market. Oil continues to be worth more, but natural gas wells are generally more productive due to shale fracking with horizontal drilling. In 1980 the optimal subsurface structure for the production of petroleum was the sand or sandstone anticline structure as depicted in figure 7. Shale was considered a confining layer that created petroleum traps. With the advent of fracking and horizontal drilling, shale is now considered a petroleum source and natural gas is the primary target. In 1980 fracking was used to mitigate for unpredicted local conditions, such as tight sands. In 2014 it is now an intentional process. Conventional oil wells have a 30% success rate, but fracked shale with horizontal drilling dramatically cuts dry hole risk and improves the possibility of success to near 100%.

Why the change in drilling tactics from 1980 to 2014? Hydraulic fracking with directional drilling into shale formations allows for improved risk control plus a dramatic increase in petroleum value over traditional well tactics. If you've been paying attention you're going to claim that oil is worth more, about 300% more than natural gas, so why the shift to shale gas?

To understand the motivations that were driving shale gas production in early 2014 we need to consider the economy of scale factors. The typical United States oil well produces 16 B/D at $100 per barrel (mid-2014 price), or $1,600 per day. Using published production figures from North Dakota, the typical natural gas well produces 560 BOE/D. At $30 per BOE the typical North Dakota shale gas well creates $16,800 per day. United States shale gas wells are creating enough natural gas to bring down the world price.

Fracking is controversial because it has the potential to contaminate drinking water supplies plus other environmental damage. There is a potential for contamination of ground water resources from the methane released by the fracking process that seeps into the water table from below.

A typical conventional oil well is drilled to a depth not to exceed 15,000 feet. Wells which are drilled horizontally, into shale deposits, are often drilled to depths at or below 3,000 feet. Aquifers from which drinking water can be extracted are generally found to a maximum depth of 1,500 feet.

A second issue of concern is the storage and transport of fracking fluid that can contaminate the water table from above. Tanker trucks are used to transport fracking fluid.

Even with the safest of drivers, accidents can occur which can result in spillage and the spreading of contaminated water.

Fracking, especially high-volume fracking repeatedly in the same rock strata, has a potential to cause minor seismic disruptions. Seismic disruptions can trigger movement along fault lines, with the potential to cause widespread damage in heavily populated areas. But the major source of fracking related seismic activity is not directly related to the primary hydraulic event.

Fracking fluid should not be mixed with surface or ground water that has the potential to be used as drinking water or for irrigation. The common solution for the disposal of used fracking fluid is to inject it under pressure into deep formations below aquifer depths. Often these formations are loaded with more fracking fluids than nature can allow them to hold. Under these conditions the fluids have the potential to migrate out of the formation by way of natural faults and fractures that exist within the rock. The high pressure injection of fracking fluids into deep formations has the potential to displace and lubricate faults that would otherwise be dormant. The result is minor seismic events. The evidence of this possibility is the fact that fracking areas are experiencing an increase in seismic activity. When operators claim that fracking isn't causing earthquakes there's a strong possibility they are technically correct. It's

49

the high pressure injection of used fracking fluid into deep rock formations that is most likely causing the seismic activity, not the fracking process.

## *Natural Gas to Liquids*

Natural gas is shipped as a compressed gas within pipelines or as LNG by specially equipped ships throughout the world. While LNG is a mature and effective technology, it requires cryogenic processing and storage systems that complicate the distribution of natural gas products because conversion equipment that is rare, complex, and expensive is required to make the natural gas useful as a fuel or feedstock product.

There is an alternative to pipeline or LNG shipment. Natural gas can be converted into a liquid that can be stored or shipped in barrels or tanks similar to crude oil. The value of this capability cannot be understated, as this is the cheapest and safest method for petroleum storage and transport. The Gas-To-Liquids (GTL) conversion is a procedure called the Fischer-Tropsch Process that will be revisited later in the book.

Inputs to the first stage of the Fischer-Tropsch GTL process are methane gas plus oxygen for the creation of hydrogen and carbon monoxide, otherwise known as Synthesis Gas or SynGas. The second stage of the process is to combine SynGas with water for the creation of diesel, naptha, or wax that can be used in their native form, or as feedstock for the creation of fuels, solvents, lubricants, plastics, or composites. Naptha, for example, is a liquid hydrocarbon fuel that is commonly used as a feedstock for the creation of high octane gasoline.

The creation of GTL plants throughout the United States to convert our abundant natural gas resources to petroleum liquids seems like a great idea, especially when you consider that the natural gas that enters the process is valued at $30 per BOE and has the potential to exit the GTL process valued at over $100 per barrel. The downside to GTL plant development is the fact that it typically takes an investment of $15-billion to create a facility for the production of 150,000 BOE/D. If you do the math you will realize that in spite of the extreme cost, a GTL plant could pay for itself in less than five years at early 2014 petroleum price factors. How would you respond if your community had an opportunity to serve the nation with a GTL plant?

# 4

## *Other Petroleum Sources*

So far we've learned about basic petroleum genesis, the traditional petroleum exploration process as taught in the 1980's, natural gas, shale gas production by use of fracking and horizontal drilling, and the natural gas to liquids process by use of the Fischer-Tropsch process. Now it's time to learn even more about non-traditional petroleum sources that are available within the North America continent and the United States in great abundance.

## Oil Shale

A better term for oil shale would be "Kerogen Shale" as the organic hydrocarbon constituents within oil shale are generally not mature. They are almost oil, or kerogen, found in a shale source bed that has not migrated into a sand reservoir yet.

There are 325-billion barrels of recoverable oil from shale formations throughout the world. The United States has 58-billion barrels of recoverable oil from shale, or 17% of the world supply.

The extraction of oil from shale has been studied by the petroleum industry since early in the 20th Century. The most promising shale oil extraction method that exists today appears to be the "In-Situ Process", developed by the Shell Oil Company.

The extraction of oil from shale by use the "In-Situ Process" is accomplished by drilling into the oil shale and placing electric resistance heaters in a hexagonal matrix throughout the formation. In this process the formation will be heated to 343°C (650°F) which is significantly hotter than a natural petroleum reservoir.

If you remember back to first chapter of this book the peak temperature for petroleum genesis is 100°C (212°F). However, the "In-Situ Process" will cook the kerogen within the oil shale for only four years, instead of the millions of years that would be used by natural processes. If the process is successful kerogen will convert to oil within the rock and migrate to where it can be extracted by traditional oil well operations. Shell claims the process is viable so long as the price of oil stays above $30 per barrel.

## Tar Sands

A better term for tar sands would be "Bituminous Soil" as the tar is actually is a variety of mature and immature hydrocarbons, to include kerogen, tar, or other petroleum liquids; and the sediment includes a variety of textures that includes sand, or clay. Tar sands are shallow, near surface deposits, found at diagenic depths and temperatures below 50°C (122°F), within the top 2,500 feet of the sediment column. You might remember that diagenesis is described in second chapter of the book as the low temperature reactions that begin to remove oxygen, sulfur, and nitrogen from organic chemicals to create immature hydrocarbon chains called kerogen, or its heavy equivalent, bitumen.

The world estimate for crude oil from tar sands is 3.6-trillion barrels, of which 2.5-trillion barrels are believed to exist in Canada, specifically from the province of Alberta. Canadian tar sand oil is exported to the United States via the Keystone Pipeline System. Three phases of the pipeline system are completed to date for the transport of 700,000

B/D or 255-million barrels per year. The fourth phase of the Keystone Pipeline System, called Keystone XL, would increase tar sand oil to the United States to 1.1-million barrels per day, or 401.5-million barrels per year. The United States dearly needs to import tar sand oil from Canada if it is to reduce its dependence on foreign petroleum, but the project is opposed by a loud minority that is working against the will of the American people.

Steam Assisted Gravity Drainage (SAGD) is considered the most effective method for extracting petroleum from tar sands. The SAGD process uses two parallel horizontal wells. The upper well is used to inject steam into the tar sand formation in order to stimulate the flow of bitumen to the lower well. The lower well extracts petroleum by traditional means. According to a 2012 report from the Canadian Energy Research Institute, an agency of the Province of Alberta, the SAGD process costs $47.50 per barrel.

The SAGD process produces bitumen which is partially refined locally in Alberta by a process called catalytic cracking that creates heavy crude, light crude, natural gas, sulfur gas compounds, and nitrous oxide. The light and heavy crude are termed SynCrude because they are a product of a refining process. By use of the Keystone Pipeline System, Canada serves as a resource provider to United States refineries along the Texas Gulf Coast. The Texas location is more ideal for refining Canadian SynCrude into finished petroleum products because it is significantly warmer than the near Arctic temperatures of the Athabasca region of Alberta.

Think of the term "SynCrude Synergy". Canada provides a raw resource to the United States where value added refinery processes are applied to increase the value of the product. Both nations benefit from the trade relationship.

# 5

## *Coal*

Earlier in the book it was explained how carbon dioxide was not a pollutant and it's appropriate to be of no particular concern as to how much carbon dioxide is produced by any particular process of mankind. However, there are dangerous gases released by the burning of some fossil fuels, although carbon dioxide is not one of them, many of these gases are produced by the burning of coal. Coal is controversial because it's a major source of acid rain and mercury that impacts water quality hundreds of miles remote from the smokestack from which it's discharged.

The United States is blessed with tremendous coal reserves, the largest in the world. It would be a tragedy to consider the magnitude of the United States coal reserves if there was no option for the use of the coal as a clean energy source. The good news is there is a clean option for the use of coal to reduce our demand on imported petroleum.

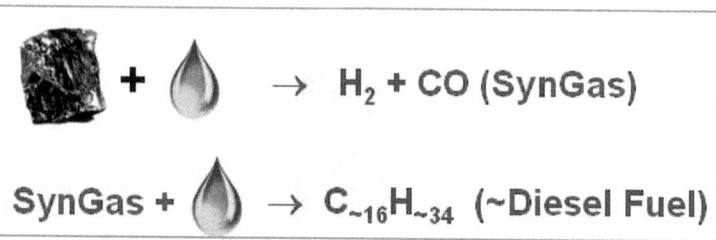

**Figure 8. Fischer-Tropsch Process for converting Coal to Diesel Fuel**

The Fischer-Tropsch process referenced earlier in the book was developed in Germany in 1919 and patented in 1923. It's how Germany fueled the Wehrmacht, Luftwaffe, and Kriegsmarine in World War II. Nazi Germany had no other choice. Surrounded by enemies they created, and with very few indigenous petroleum supplies, Germany had abundant coal and needed a way to convert it into liquid fuels. Coal can be converted into SynGas, and the SynGas can be converted into petroleum liquids such as diesel fuel as illustrated in figure 8.

In the first stage of the Fischer-Tropsch Process for coal, SynGas is created from coal plus water to create hydrogen plus carbon monoxide. In the second stage of the process SynGas plus water is converted to synthetic liquid petroleum very similar to the composition of diesel fuel. The diesel product can be used as a fuel or as a feedstock for other petroleum based products. The trucking industry, railroads, jet aircraft, and the military would greatly benefit from an increase in indigenous diesel supplies created from within the United States.

One ton of coal has the potential to produce 2.2 BOE of liquid petroleum. The United States has the largest known coal reserves in the world with 237 billion tons, which equates to 522 billion BOE, or twice the known petroleum reserves of Saudi Arabia. If a national policy to use coal and natural gas for the production of liquid petroleum by use of synthetic processes could be established, the United States could once again lead the world in petroleum resource potential based on the immense supply of coal and natural gas reserves. The United States has the potential to have a 286 year supply of synthetic petroleum from coal at a production rate of 5-million BOE/D.

Many nations and the United States Department of Defense are looking toward synthetic petroleum created from coal as a liquid fuel energy alternative. South Africa, Malaysia, Qatar and Finland have synthetic Fischer-Tropsch petroleum plants in operation today. There are small scale Fischer-Tropsch demonstration plants in the United States. The United States Air Force is currently in the final stages of testing and certifying the use of synthetic JP-8 fuels for use in its aircraft with the potential that these fuels will be of superior quality with improved air quality effects.

Consider the economics of Fischer-Tropsch conversion of coal to liquid petroleum products. If a producer pays $85 per ton for coal, the basic cost of the liquid petroleum product would be $39 per BOE. However, Coal-To-Liquid (CTL) technology is very expensive. According to the latest published figures, a 50,000 BOE/D CTL plant

*The world estimate for crude oil from tar sands is 3.6-trillion barrels.*

would cost $5-billion to build, and the cost of the product would be $70 per BOE when the cost of the conversion plant has to be considered. Buts that's still significantly cheaper than the mid-2014 price of $100 per barrel for oil. The nation of South Africa currently leads the world in CTL fuel production with a capacity of 180,000 BOE/D.

# 6

## *The Case for Fischer-Tropsch*

Now that we know more about the potential for natural gas conversion to synthetic petroleum liquids, let's consider a case for how it can used to optimize long range investment outcomes. In the second chapter we learned about the Cove Point LNG Plant that is working toward an expansion project for the export of natural gas. The current project plan calls for the expenditure of $3.5-billion in order to convert an existing LNG import facility into an LNG export facility. Instead of using the $3.5-billion for the development of an LNG export plant, let's consider the possibility of building a Fischer-Tropsch GTL plant instead.

The goal of this analysis activity is to determine if there is more income potential for the Cove Point Plant if they build a Fischer-Tropsch GTL plant instead of an LNG plant.

Column B of figure 9 lists the value factors for the LNG export facility based on best available information released to the public. The analysis model is calculated from the volume of petroleum that can be transported using the 500,000 barrel capacity of Panamax oil tankers and LNG tankers of similar BOE volume. The 20 year gross value of the products shipped includes a subtraction for the cost of the LNG or GTL export facilities. There is no way to calculate the net value of these project concepts from information available to the public.

| A | B | C | D |
|---|---|---|---|
| PROJECT ANALYSIS FACTORS | LNG EXPORT FACILITY | FISCHER-TROPSCH GTL EXPORT FACILITY OPTION 1 | FISCHER-TROPSCH GTL EXPORT FACILITY OPTION 2 |
| Cost of Plant | $3.5-Billion | $3.5-Billion | $11-Billion |
| Daily Capacity | 160,000 BOE | 35,000 BOE | 110,000 BOE |
| Days to Load 500,000 B/BOE Ship | 3.1 | 14.3 | 4.5 |
| Number of Ships per Year | 80 | 24 | 80 |
| Annual Production Capacity | 40-Million BOE | 12.2-Million BOE | 40-Million BOE |
| Annual Product Value | $1.2-Billion ($30/BOE) | $1.22-Billion ($100/BOE) | $4-Billion ($100/BOE) |
| 20-Year Gross Product Value | $20.5-Billion | $20.8-Billion | $69-Billion |

**Figure 9.  Analysis of LNG vs. Fischer-Tropsch GTL Export Facilities**

Column C describes the Option 1 value factors for a hypothetical GTL plant of similar cost to the LNG plant that is currently under construction. The Option 1 GTL plant has significantly less daily production capacity than an LNG plant of the same cost because of the reduced production capacity. The GTL plant will handle only 24 ships per year but the GTL product has a higher value. This analysis model will place the value of the GTL synthetic petroleum liquid at $100 per BOE, and the LNG value is placed at $30 per BOE, the approximate values of crude oil and natural gas in mid-2014. Even though the Option 1 GTL plant will handle 56 fewer ships each year, the gross value of the exported product is worth essentially the same as the LNG plant product.

*The United States is blessed with tremendous coal reserves, the largest in the world.*

The value factors for a second option related to a Fischer-Tropsch GTL plant is listed in column D. The GTL Option 2 plant is scaled to produce synthetic petroleum liquids to match the BOE production of the LNG plant in column B. Although the cost of the GTL Option 2 plant is triple the cost of the LNG plant, the gross 20-year value of the synthetic petroleum liquids produced by this GTL plant would be worth $48.5-billion more than LNG products over that time span.

There are other factors to consider that make Fischer-Tropsch GTL plants more attractive besides the product value. The primary GTL plant product is a synthetic petroleum similar to diesel fuel that is far safer and cheaper to store and transport than LNG. Although diesel fuel will burn, it will not explode like natural gas. Another significant advantage of synthetic petroleum liquids is that they can be shipped by traditional crude oil tankers. Oil tankers cost about $90 per barrel capacity to build verse the cost of LNG tankers that cost about $230 per BOE. The transit fees and insurance costs for oil tankers are also significantly cheaper than LNG tankers.

We just completed a short, back of the envelope analysis that suggests a Fischer-Tropsch GTL plant is more profitable and safer for the local community it serves within than a traditional LNG approach. Let's take our Fischer-Tropsch based solution one step farther. The Cove Point LNG Plant is connected to the gas fields of West Virginia and Pennsylvania by an existing pipeline. Instead of putting natural gas in the pipeline, let's consider placing SynGas created from the first stage of the Fischer-Tropsch CTL we learned about in the previous chapter. The shale gas potential from the Allegany region is a new phenomenon, coal is the legacy product for this region. Fischer-Tropsch CTL technology would give the region a clean coal option that would allow the United States to dominate the world oil market with synthetic petroleum liquids.

Fischer-Tropsch GTL or CTL is a two stage process. Stage one of the process is the conversion of natural gas or coal plus water into hydrogen gas plus carbon monoxide, called SynGas. Let's consider placing a series of stage one SynGas plants throughout the West Virginia and Pennsylvania coal and natural gas region. This same region is served by existing natural gas pipelines. Instead of using the pipelines for natural gas,

*One ton of coal has the potential to produce 2.2 BOE of liquid petroleum.*

let's use them to transport SynGas to coastal export facilities. Instead of LNG plants at the coastal facilities, let's perform the Fischer-Tropsch stage 2 process; the creation of diesel fuel from SynGas.

The creation of synthetic petroleum liquids from coal and natural gas is the best possible solution to solving this nation's dependence on foreign oil. The technology is mature and with the appropriate economy of scale it will create more wealth than legacy coal or LNG distribution with reduced risk on the environment. Fischer-Tropsch processing plants should be a part of the nation's strategy to end our dependence on foreign oil.

The diesel fuel created from Fischer-Tropsch conversion of coal and natural gas would be vital to the United States energy economy. New sources of diesel fuel derivative products from domestic producers will reduce the cost of operations for the trucking industry, railroads, jet aircraft, and the military.

# 7

## *Review of Petroleum Sources*

In this next chapter we'll review the petroleum sources available from within the United States and work toward a hypothetical concept for energy independence within ten years from the year 2014. United States petroleum production and consumption for year 2013 is listed below.

|  | **2013** |
|---|---|
| **OIL PRODUCTION (B/D)** | 7,535,000 |
| **OIL CONSUMPTION (B/D)** | 18,490,000 |
|  | (10,955,000) |
|  |  |
| **NATURAL GAS PRODUCTION (BOE/D)** | 13,814,000 |
| **NATURAL GAS CONSUMPTION (BOE/D)** | 14,808,000 |
|  | (994,000) |

**Figure 10. US Petroleum Production and Consumption for Year 2013**

According to figure 10 the United States imports close to 11-million B/D more oil than it produced in the year 2013. Even natural gas had a supply deficit of close to 1-million BOE/D in year 2013, in spite of the fact the United States is the number one natural gas producer in the world.

Now let's take a hypothetical look at the year 2024, ten years into the future as reflected in figure 11. The oil and natural gas production figures for year 2024 were compiled from projections calculated and published by the United States federal government via the Energy Information Agency. The oil and natural gas consumption figures match the values for year 2013 because Energy Information Agency projections foresee no significant increase in petroleum demand based on the concept that automobiles will continue to get more fuel efficient and alternative energy sources such as wind and solar power will reduce the need for petroleum based electricity power generation. Even with the production increases projected by the federal government, it appears that there will be an oil deficit of close to 8-million B/D by year 2024. However, federal government projections suggest that natural gas could have a significant surplus of 3.4-million BOE/D by 2024.

| | **2024** |
|---|---:|
| **OIL PRODUCTION (B/D)** | 10,500,000 |
| **OIL CONSUMPTION (B/D)** | 18,490,000 |
| | (7,990,000) |
| | |
| **NATURAL GAS PRODUCTION (BOE/D)** | 18,200,000 |
| **NATURAL GAS CONSUMPTION (BOE/D)** | 14,808,000 |
| | 3,392,000 |

**Figure 11. US Petroleum Production and Consumption Projections for Year 2024**

How can the United States be petroleum energy independent by year 2024? Figure 12 provides the answer. The United States can meet the demand for 18.5-miilion B/D or BOE/D consumption from five sources of petroleum already described in this book to include oil, tar sand SynCrude, oil shale, and the Fischer-Tropsch conversion of surplus natural gas and coal to synthetic petroleum liquids via GTL and CTL processes.

This energy independence model is based on the concept that oil shale would be extracted at a rate of 1% of available reserves per year, or a 100 year supply. Coal extraction is based on the concept that 0.1% of reserves would be used per year, or a 1,000 year supply.

However, the portion of the petroleum liquids from the natural gas surplus and coal will take a significant investment. The year 2024 was selected for this hypothetical concept because it will take at least ten years of political activity and engineering studies to make the possibility of energy independence a reality.

| | **2024** | |
|---|---|---|
| | 10,500,000 | OIL |
| From Natural Gas Surplus | 3,392,000 | GTL |
| From Keystone XL | 1,100,000 | Tar Sand SynCrude |
| If 1% of Reserves Used Each Year | 1,600,000 | Oil Shale |
| If 0.1% of Reserves Used Each Year | 1,900,000 | CTL (Coal) |
| | 18,492,000 | |

**Figure 12. US Energy Independence from Five Sources for Year 2024**

Fischer-Tropsch GTL/CTL plants cost about $100,000 for each BOE/D of capacity. Petroleum independence by the conversion of surplus natural gas and coal to meet the need for the production of 5.3-million BOE/D would require an investment of approximately $530-billion. In an ideal world the cost of Fischer-Tropsch technology should not be a burden to the United States taxpayer. The proper role of government in this petroleum independence venture should be as a regulatory facilitator in order to encourage private investment for the creation of new wealth, expand the economy, and provide energy security long into the future. Outside of basic discovery and invention research and development, the government should not be a funding source. The United States should be energy independent without expanding the size of government.

But there's more!

# 8

## *Methane Hydrate*

Methane Hydrate is methane ($CH_4$) trapped inside a lattice of frozen water molecules ($CH_4 \bullet 5.75\ H_2O$). It exists in nature where temperatures are below the freezing point of water and the pressure exceeds 5 MPa, or megapascals, which equates to 50 atmospheres of pressure, or 725 lbs/in$^2$.

Methane is typically a gas, but exists as a solid mixed with ice when it's found below the freezing point of water, and buried below 1,500 feet in depth. There is no temperature-pressure relationship where methane exists as a liquid. When methane passes from a solid to a gas it expands by a factor of 164 times its solid volume. This expansion has explosive potential. Without flame, spark, or exposure to oxygen, frozen methane will convert to its gas form by simply passing a temperature-pressure boundary. Methane hydrate will not be viable as a hydrocarbon fuel source until extraction technology is developed that can safely bring it to the earth's surface without the danger of explosion.

Methane hydrate is found below the deep ocean seafloor, on the seafloor at fault seeps, and within Arctic permafrost deposits. There is significant methane hydrate deposits found throughout the world along the margins of every continent. Methane hydrate discovery and exploitation as a hydrocarbon source is the next petroleum frontier.

> *Synthetic petroleum liquids from coal and natural gas is the best possible solution to solving this nation's dependence on foreign oil.*

Massive deposits of methane hydrate have been mapped off the coast of the Carolinas. United States Geologic Survey calculations of the volume of these hydrocarbon deposits are estimated at 222-billion BOE. To understand the magnitude of this volume of petroleum we need to consider these deposits in reference to the Saudi Arabia Ghawar Oil Field we learned about earlier in the book. Let's create a new unit of measure called a SAGOF, which will represent the estimated volume of petroleum reserves remaining within the Saudi Arabia Ghawar Oil Field, or 100 billion barrels, or BOE. The off-shore Carolina methane hydrate deposits are estimated to be 2.2 SAGOFs. That's a significant volume of petroleum!

The Arctic region, which includes the nations of Russia, Canada, the United States, and Norway, has the greatest volume of methane hydrates in the world because the colder temperatures of the Arctic allow for these deposits to exist at shallower depths. Many studies have been conducted to estimate the volume of methane hydrate in the Arctic with a midrange value on the order of 300 SAGOFs.

But the most compelling region of the world that is known to have methane hydrate deposits is the nation of Japan, within the Nankai Trough. A low estimate for the volume of methane hydrates within the Nankai Trough is measured at 60 SAGOFs. Japan has a joint venture with the United States Geologic Survey for the discovery and development of production technology for the safe and effective extraction of methane hydrate from deep ocean deposits. How big is 60 SAGOFs of natural gas? Sixty SAGOFs equals a 3,650 year supply of petroleum at Japan's current consumption rate of 4.5-million B/D.

Japan is the world's major consumer of petroleum that does not have an indigenous supply. Couple this fact with the notion that Japan is a technologically advanced nation that has a strong motivation to solve its energy supply problems. Japan is the nation that is most likely to lead the world toward safe and effective technologies for extracting methane hydrate deposits.

Japan has a strong ethos for the development of visionary technological systems and processes. It's not hard to imagine the possibility that within the next decade or so, large platform communities will hover over the deep ocean, extract methane hydrate deposits, and convert the natural gas to liquid fuels via integrated Fischer-Tropsch GTL conversion plants. The potential exists that Japan could do more than just meet its domestic demand for energy. An indigenous supply of methane hydrate could allow Japan to become an exporter of petroleum to the rest of the world. What will happen to the world price when Japan stops importing mid-east petroleum?

# 9

## *United States Petroleum History*

For the first one hundred years after the discovery of the Pennsylvania Drake Well in 1859, the United States was the dominant oil producer in the world. The Second World War was fought with our own domestic oil production and there was still enough surplus oil to export to our allies. Although those days appear to be behind us a national commitment toward petroleum independence could reverse the trend. It's very important to know our petroleum history with a goal in mind to correct for the errors of the past in order to prepare for petroleum dominance again in the near future.

## 1900's: First Mid-East Discovery

At the turn of the Century United States petroleum consumption reached 200,000 B/D, all of which was produced domestically. On January 10, 1901, Anthony F.

Lucas struck oil at 1,139 feet at Spindletop Hill, a few miles south of Beaumont, Texas. Spindletop was the largest oil discovery in the world up to that time and soon produced over 100,000 barrels per day. As a result of the discovery over two hundred new oil companies formed to explore and produce petroleum throughout the region including two well-known companies, Gulf and Texaco. Standard Oil was specifically barred from operations within the state of Texas due to its reputation for monopolistic policies. Through three cycles of drilling activity, over 153-million barrels of oil were produced from the Spindletop Oil Field by 1985.

On the other side of the world, a geologist by the name of George Bernard Reynolds struck the first discovery well in the Persian Gulf region in May of 1908, after seven years of dry holes. With funds exhausted, the oil discovery was accomplished after transmission of a telegram from his employer, an Englishman by the name of William Knox D'Arcy, directing him to *"cease work, dismiss the staff, dismantle anything worth the cost of transporting to the coast for re-shipment, and come home."* His persistence paid off. The result of his effort led to the creation of the Anglo-Persian Oil Company, a major international corporation later called British Petroleum, which today is known as BP. Iran continues to be a major producer of petroleum, with proven reserves of 154.6-billion barrels, number four in the world. This first discovery would lead to many more throughout

Iran and the broader region so that five Persian Gulf nations control 56% of the world's known petroleum reserves.

## *1910's: Break-Up of Standard Oil*

By 1910 United States petroleum production was 600,000 B/D, and the majority of it was produced by a single company called the Standard Oil Company. John D. Rockefeller, the founder of Standard Oil, entered the petroleum industry when it was a new and emerging sector of the economy in the 1860's. He followed the practice of other robber barons of the 19th Century by buying out his competitors. Ultimately he developed a vertical and horizontal monopoly that controlled 90% of the United States petroleum industry that included oil wells, pipelines, refineries, and retail distribution infrastructure. The major positive contributions of the Standard Oil Company was the development of efficient refining processes that reduced the cost of production and resulted in a variety of petroleum products with high quality "standards" that were safe, reliable, and predictable within the commercial market place. The negatives for Standard Oil Company practices were the fact that it controlled too much of the United States petroleum market and used its power and influence to constrain competition and interstate trade.

Rockefeller manipulated petroleum prices to put his competitors out of business, and then raised prices higher than the previous market after the competition was gone. He also failed to distribute petroleum to all customers equally by giving low prices to some sectors of the economy, such as the railroad industry, and applied higher prices to other customers. The Sherman Antitrust Act, passed by Congress in 1890, was used by the Supreme Court in 1911 to break-up the Standard Oil Company into 34 smaller companies throughout the

*Sixty SAGOFs equals a 3,650 year supply of petroleum at Japan's current consumption rate.*

United States and other foreign countries. However, the United States would not have been ready for the advent of the great number of automobiles that would enter the commercial marketplace after the introduction of the Model T Ford in 1908 without the petroleum infrastructure developed by Rockefeller and the Standard Oil Company.

## 1920's: Teapot Dome Scandal

By 1920 the United States appetite for petroleum had grown to 1.4-million B/D. It wasn't just the automobile industry that had a growing thirst for petroleum fuel, the United States Navy made a conversion from coal fired to fuel oil ships early in the 20th Century. A section of Wyoming, called the Teapot Dome, and California, called the Elk Hills, known to have petroleum, were designated as Naval Petroleum Reserves and set as off-limits to exploration and development unless a strategic national emergency was declared. Control of the reserves was transferred from the US Navy to the Department of the Interior early in the Warren G. Harding administration. Although President Harding had the good fortune to die in office before he could be implicated, his Secretary of the Interior, Albert Fall, was prosecuted by the federal government and convicted of accepting bribes for arranging secret deals with Edward Doheny and Harry Sinclair to drill within the Naval Petroleum Reserves of Wyoming and California.

## 1930's: Great Depression

Petroleum production reached 2.25-million B/D in the year 1930. The good news for oil occurred in October 1930 when a wildcatter by the name of "Dad" Joiner struck oil near Tyler, Texas. Joiner made what was to become the largest discovery since Spindletop. The bad news was that the Great Depression hit coincident with his strike and the price of oil dropped to 10 cents per barrel in 1931. The East Texas Oil Field was the largest discovered in the lower 48 states, 45 miles long by 5 miles wide. With 30,340 wells drilled, the field would ultimately yield 5.4-billion barrels of oil. To keep this fact in context, the East Texas Oil Field is enormous by United States standards, but less than 5% of what the Ghawar Oil Field of Saudi Arabia will eventually produce. The East Texas Oil Field discovery was critical to United States victory in the Second World War.

## 1940's: World War II

The United States military constructed 6,768 ships, 23,113 armored fighting vehicles, and 300,000 aircraft to defeat Germany, Japan, Italy and their minor allies. To achieve this goal the United States produced 6.1-billion barrels of oil from 1942 through 1945, or 4.5-million B/D, all of which was produced domestically, primarily from the states of Texas, Louisiana, Oklahoma, Pennsylvania and California.

The United States dominated oil production during World War II as depicted in figure 13. The graphic illustrates oil production for the major nations at war for the year 1943, as this is the only year for which there is complete information. The Allies controlled the lion's share of petroleum, or 93.6% of the World War II production. British Empire petroleum was primarily produced in Iran by the Anglo-Persian Oil Company.

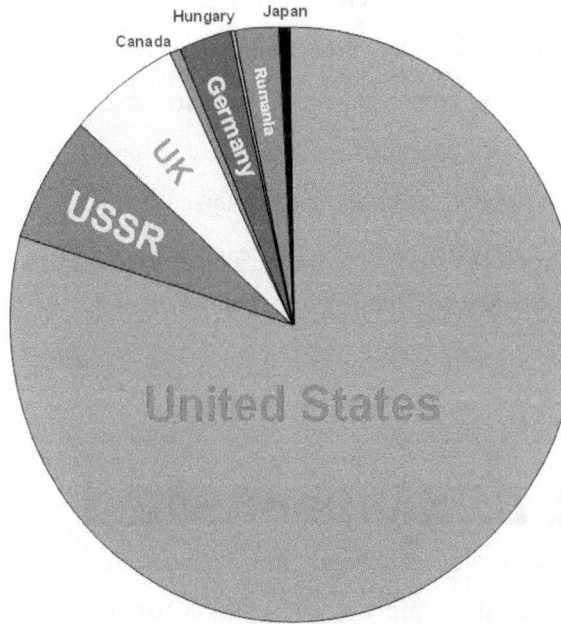

| NATION | 1943 |
|--------|------|
| U.S.A. | 80% |
| USSR | 7.2% |
| UK | 6.3% |
| Canada | 0.5% |
| Germany* | 3.0% |
| Italy | 0.004% |
| Hungary | 0.3% |
| Rumania | 2.1% |
| Japan | 0.9% |

**1943 World Production =
1,837,704,300 Barrels**

**\* 65% of German Petroleum
from Fischer-Tropsch CTL
(36,212,400 barrels)**

**Figure 13. Petroleum Production Pie Graph for World War II, 1943.**

The Axis controlled a humble 6.4% of the petroleum used in World War II. Sixty-five percent of German petroleum was created from Fischer-Tropsch CTL plants. The capture of petroleum resources was a strong motivating factor that compelled the Axis powers to enter World War II. Japan's primary source for petroleum was the capture of oil fields within Malaysia, Borneo, and the Dutch East Indies. German attempts to capture Soviet oil fields were not successful. The Maikop oil fields, captured by Germany in August 1942, were destroyed by retreating Soviets, and never brought into production to support the Wehrmacht.

The Grozny and Baku oil fields were targeted as strategic objectives but never reached by the German Army. A contributing factor to the ironic failure to capture these remote fields was an over extended supply line combined with a lack of oil needed to support a mechanized army. Germany didn't have enough petroleum to capture more. Japan had enough petroleum to capture more, but not enough to keep it.

## 1950's: Shift To The Middle-East

In the post-war economic expansion the demand for petroleum in the United States grew to 6.57-million B/D by 1950. Domestic petroleum sources in the United States appeared unable to meet increasing demand and the nation looked to the Middle-East as a source for imports. Mid-east oil was cheap, plentiful and seemed like a stable source of petroleum for the foreseeable future. This strategy was adopted before the advent of Radical Islam. In 1950 a mere 350,000 B/D of petroleum were imported into the United States. Although it was a modest volume at the time, the trend was established.

## Seven Sisters

Arab oil nations were not the first to impose an embargo related to mid-east oil. That claim belongs to the British who controlled the great majority of Persian Gulf oil resources in 1950. As a response to nationalization of Iranian oil in 1951, the British imposed an international embargo, blockade, and withdrawal of its technical labor force from the world's largest oil refinery at Abadan, Iran. The British position was based on the belief that Iranian oil was only possible because of the investment of British risk capital, British engineering, British labor skills, and British technology. With the exception of Italy, all of the world's major powers supported the British policy. The motivation behind the international support for the embargo was the realization that if petroleum nationalization worked in Iran, it could lead toward the beginning of a trend throughout the rest of the world. In reality the actions of Iran was the beginning of a trend.

To end the stalemate, the United States Department of State in 1954 recommended the formation of a "Consortium for Iran", created from major oil companies to return Iranian oil back to the international markets. Behind the scenes, Mohammad Reza Shah Pahlavi, better known as the Shah of Iran, was given a bribe from the CIA to support a coup to oust the Prime Minister, Mohammad Mosaddeq, responsible for nationalization. The formation of the consortium was an opportunity for international oil companies outside of Britain to participate in the redevelopment of Iranian oil. The Consortium for Iran was comprised of the seven petroleum companies listed as follows, otherwise known as the Seven Sisters:

- Anglo-Persian Oil Company (BP, United Kingdom)
- Gulf Oil
- Royal Dutch Shell (Netherlands/United Kingdom)
- Standard Oil of California (Chevron)
- Standard Oil of New Jersey (Exxon)
- Standard Oil Co. of New York (Mobil)
- Texaco

Before there was OPEC, these "Seven Sisters" not only controlled the production of mid-east oil, but 85% of the oil produced throughout the world. Five of the seven oil companies were remnants of the original Standard Oil Company that was broken up into smaller companies by the United States Supreme Court in 1911.

## 1960's: Formation of OPEC

By 1960 the United States consumed 8.3-million B/D of which 950,000 B/D were imported. This is the decade where the wealth and power of the United States was at its zenith.

In spite of funding the "Marshall Plan" to re-build Europe and Japan after World War II in the previous decade; construction of the Interstate Highway System; fighting the Viet Nam War; waging a War on Poverty that created more poverty; developing Cold War nuclear weapon systems to counter the Soviets, and the Moon Program; the economy continued to grow and the United States was able to end the decade without significant debt and remain the world's major economic power.

In 1959, President Dwight Eisenhower invoked a clause from the Reciprocal Trade Act Amendments of 1955 that imposed oil import quotas. As usually happens when the government takes an action it thinks will correct a condition, the law of unintended consequences take effect. Four mid-east nations plus Venezuela that were heavily dependent on petroleum export as a major component of their economy formed the Organization of Petroleum Exporting Countries (OPEC). Shut-out from the United States market, OPEC turned to Europe for exports by lowering prices, flooding their market, and effectively increasing their dependence on cheap foreign oil.

The founding members of OPEC in 1960 were Iran, Iraq, Kuwait, Saudi Arabia and Venezuela. The nations of Qatar, Indonesia, Libya, the United Arab Emirates, and Algeria joined OPEC by the end of the decade. Seven of the original ten member nations were located in the Persian Gulf

Region. Nine of the ten nations were predominantly Muslim religion.

Organized in Baghdad, Iraq, the original OPEC headquarters was in Geneva, Switzerland, but is now located in Vienna, Austria. As of year 2012, OPEC nations claim to control 81% of the world's proven reserves, or 1.2-trillion barrels. From their charter statement, *"OPEC's objective is to co-ordinate and unify petroleum policies among Member Countries, in order to secure fair and stable prices for petroleum producers; an efficient, economic and regular supply of petroleum to consuming nations; and a fair return on capital to those investing in the industry."*

In summary, the goal of OPEC is to collectively find a standard price and production figures that maximize profits and conserve the resource, in order to create wealth for member countries without harming the economies of the customers they are supplying. In effect, OPEC wants to be the most effective parasite it can possibly be without killing its host. Petroleum consumers throughout the world can complain, but in truth OPEC uses the same tactics with the same motivations as the major international oil companies they compete against; the only difference is where the wealth ends up.

One of the mantras from OPEC that must be respected is the realization that the largest factor impacting the cost of petroleum in many parts of the world is the taxes levied by various national governments. Too often the focus related to the high cost of petroleum is placed on OPEC or the international oil companies, but in truth it is often the national government taking the lion's share; without having to take any risk in the challenges related to providing investment resources or significant contributions to the effort of exploration, development, refining, distribution, or marketing.

## 1970's: Emergence of Mid-East Politics

By 1970 the United States consumed 11.2-million B/D of which 1.75-million B/D were imported. As war was winding down in Southeast Asia, a surprise war erupted in the Middle-East. In October 1973, Egypt and Syria led an attack on Israel on Yom Kippur, the highest holy day of the Jewish calendar. Much of the Arab world and beyond contributed military forces to the effort including the nations of Cuba, Kuwait, Morocco, North Korea, Saudi Arabia, and Tunisia. The Arab led forces had initial successes in the early stages of battle, but Israel was able to call up its reserves and mobilize a defense that quickly turned into an attack, and ultimately a rout. With Israeli troops threatening Damascus to the north, and with another Israeli army racing toward Cairo to the south, the United Nations negotiated a cease fire agreement that ended combat operations.

## 1973 Oil Embargo

OPEC responded to the results of the Yom Kippur War by proclaiming price increases plus an embargo on oil shipments to Israel and its allies, to include the United States. Petroleum prices rapidly rose by 400% from $3 ($5.60) to $12 ($22) per barrel. Fuel shortages, inflation, oil price controls and recession in the United States quickly followed.

## North Sea Oil

Natural gas was first discovered in the United Kingdom sector of the North Sea in 1965. These resources were largely ignored until the price increases that resulted from the 1973 OPEC Oil Embargo made high risk off-shore platform drilling a more attractive source for petroleum. The Brent Field was discovered in the Norway sector of the North Sea in 1976. By 1980 North Sea petroleum production reached 2-million B/D, with a portion of it exported to the United States. Over 100 major on-shore and off-shore oil fields have been discovered in Norway, United Kingdom, Netherlands, Germany, and Denmark, and are now producing petroleum in a region of the world that groaned from having to import oil for the first 100 years of the petroleum age.

## Trans-Alaska Oil Pipeline

Oil was discovered in Prudhoe Bay on the north slope of Alaska in 1968. But the region is frozen nine months of the year and the possibility that oil tankers could reliably ship the resource to remote oil refineries was not realistic. Private initiative funded the project that took 3.2 years and $8-billion to complete. The only item the Trans-Alaska pipeline needed from the government was permission. The pipeline links Prudhoe Bay on the Arctic Ocean with an ice free port facility called the Valdez Terminal on Prince William Sound, 800 miles away. The project was completed in May 1977 and moves up to 2.1-million B/D. Production figures for the region have been in gradual decline since 1988, not because of a lack of petroleum, but because the region needs more government permission. The state of Alaska is generous with its oil proceeds. Alaskan citizens throughout the state don't file their taxes to determine what they owe; they file in order to determine what they will be paid from collective oil profits.

## Iranian Revolution

Harmony and trust was never fully achieved between the Shah and the Iranian people since he came to power as a result of a British and Soviet resourced coup in 1941. With first the British, and then the United States standing behind him, the Shah lived under the delusion that his reign was a modern follow-on to the legacy of the ancient Persian Empire. In reality his regime was a time bomb ready to explode. When the Shiite cleric Ruhollah Khomeini returned from exile in 1979, the Shah's government collapsed, to be replaced by an Islamic Republic based on theocratic principles. The United States lost its strongest ally among the Persian Gulf nations; all foreign owned oil assets were seized and nationalized. American citizens had no choice but to flee the country.

> *OPEC uses the same tactics with the same motivations as the major international oil companies.*

The arrest and imprisonment of United States diplomats plus the attempt to rescue them by use of a failed military operation were the critical events that doomed the 1980 re-election attempt of Jimmy Carter.

The revolution wasn't good for Iranian petroleum as production dropped from a peak of 6-million B/D in 1974 to 1.4-million B/D by 1980 as a result of having to re-organize the industry after foreign oil field and refinery technicians were driven from the country. It was a replay of the events of 1951 when the British departed Iran, only this time the US Department of State couldn't intercede; diplomatic relations between the United States and Iran were terminated in April

1980. Relations between Iran and the United States remain strained to this day. Iran was a major exporter of petroleum to the United States at that time, and the loss of this oil source created short term supply disruptions that resulted in gasoline shortages, price increases, long lines, and ultimately behavior modifications that had a significant impact on petroleum sources and issues in the following decade.

## 1980's: Oil Glut

Americans entered the new decade concerned about the high cost of energy, inflation, and continued turmoil in the Middle-East. The high cost of gasoline generated a collective response that had a positive impact on long term trends. The response from consumers improved United States automobile fuel mileage efficiency that entered the decade at 16 MPG average and increased it to 21 MPG by the end of the decade, in spite of the fact that Reagan dropped federally imposed efficiency standards. In the first two years of the decade Americans drove less too, by collectively driving 50-billion fewer miles in 1981 than they did in 1979.

United States petroleum production actually increased in the second half of the 1970's so that by year 1980 total US oil production reached 8.6-million B/D. Of that total the Alaska North Slope produced 1.6-million B/D, off-shore platforms in federally controlled Gulf of Mexico waters produced 600,000 B/D, and domestic production from the lower 48 states reached 6.4-million B/D. Much of the domestic production was the result of private initiative from thousands of small, independent oil companies that were responding to high oil prices that seemed like they would continue to increase long into the future. Total United States demand reached 17.1-million B/D by 1980, of which 5.3-million B/D were imported; primarily from Algeria, Saudi Arabia, Nigeria, Mexico, Canada, and the United Kingdom. Yes really . . . the United Kingdom was an oil exporter based on the success of its North Sea oil platforms!

## Iran-Iraq War

In September 1980 Iraq declared war on the new Iranian republic. It's beyond the scope of this book to explain the complex set of motivations behind the invasion, but a primary military objective for Saddam Hussein was uncontested control of the Shat al-Arab waterway and a connecting port facility such as the existing refinery at Abadan, Iran on the Persian Gulf. Besides the horrendous loss of life, property, and equipment, the war denied both countries access to the waterway and reduced the ability of each to export oil to the rest of the world. From a peak of 3.5-million B/D in 1978, Iraq production dropped to 1.0-million B/D by 1981. Iran oil production dropped to 1.2-

million B/D as a result of the war. Iraq entered the war as a virtual US ally but this condition changed in May 1987 when an Iraqi Mirage F1 fighter struck the USS Stark FFG with two Exocet cruise missiles. This was not an isolated event; over one hundred merchant tankers from all parts of the world were damaged or destroyed by cruise missiles launched from both of the belligerents. The war ended in a stalemate in 1988; neither nation could claim victory, with broken economies, alienated from the rest of the world. Although the war was effective at creating supply disruptions related to oil shipped from all the Persian Gulf nations, production decreases were more than matched by increases from Mexico, Nigeria, Venezuela, Soviet Union, and the North Slope of Alaska.

## Oil Price De-Regulation

Jimmy Carter had a process in place to remove the price controls imposed in 1973 on domestic oil by a series of phased steps. Ronald Reagan removed them in his first executive order by the stroke of his pen in one grand move, in his first week in office. Domestic independent oil drillers were surprised by the result. Expecting oil prices to increase because they thought the government controls were keeping prices lower than the real market value; they found out that government controls were keeping prices higher than the real market value.

The price of oil dropped dramatically after price de-regulation took effect coincident with driver choices toward smaller, more fuel efficient cars, and reduced driving miles in the United States and much of the developed world. Many independent drillers were caught over-exposed in well ventures that required increasing oil prices to justify the cost-risk models they were operating under. Oil prices decreased from a high of over $37 ($106) per barrel in 1980 to under $15 ($31) per barrel by 1986. Thousands of these small oil companies went bankrupt or re-organized their assets to escape losses. Low oil prices were great for the expansion of the economy throughout the 1980's, but domestic oil production dropped in the lower 48 states throughout the decade in deference to cheaper oil imported from foreign sources.

## 1990's: Mergers

No one that was an adult in the early 1980's had an imagination for how, if, or when the Berlin Wall would fall; the unification of East and West Germany; or the collapse of the Soviet Union. From the perspective of history, these events happened in rapid succession as the Cold War ended without violence or bloodshed; a force greater than man was at work. The Berlin Wall fell in November 1989, followed by the fall of the Soviet Union in December, and the two Germanys re-united in October 1990. As the decade unfolded we watched a giant dysfunctional Soviet empire disintegrate and re-emerge as a smaller dysfunctional nation of Russia.

For the first time in the age of petroleum the United States entered a new decade with oil consumption lower then than the decade before. Oil consumption dropped 100,000 B/D lower, to 17.0-million B/D, of which 5.9-million B/D were imported; primarily from Saudi Arabia, Nigeria, Mexico, Venezuela, Canada, Iraq, Angola, United Kingdom, Columbia, Kuwait, and Algeria.

## Gulf War I

Iraqi oil production quickly recovered after the end of their recent war with Iran just two years before so that it reached 2.9-million B/D by 1990. But Saddam Hussein wasn't satisfied. Unrestricted access to the Persian Gulf couldn't be achieved from Iran so he seized it from a weaker neighbor, Kuwait, in August 1991. International support for the war was so strong that even the United Nations Security Council could ratify a resolution to remove Iraq from Kuwait by military force. The air war began in January 1992, followed by a 4-day ground war in late February. George Bush I did a masterful job leading the effort to liberate Kuwait with force contributions from the United Kingdom, France, Saudi Arabia, Syria, and Egypt. The American people repaid his effort by not re-electing him in 1992.

With all this history in the news, petroleum related events were missed by the general public. Many of the remnants of the original Standard Oil Company that were dismantled by the Supreme Court in 1911, reformed:

- The original Anglo-Persian Oil Company that pioneered exploration and development of mid-east oil over 100 years before; changed its name to British Petroleum after returning to Iran as a member of the "Consortium of Iran" in 1954. British Petroleum bought Standard Oil of Ohio, known to the public as Amoco, in 1987, re-organizing under the name BP. Other notable acquisitions for BP include Atlantic Richfield, Castrol, and Aral, a European petroleum company.

- Standard Oil of California, better known to the public as Chevron, bought Gulf Oil Company in 1984, followed up by the purchase of Texaco in year 2000, making it the second largest petroleum company in the United States.

- Standard Oil of New Jersey, known as Exxon since 1972, merged with Standard Oil of New York, known as Mobil, to form the largest international oil company in the world, ExxonMobil, in 1999.

The three oil companies referenced above plus Royal Dutch Shell comprise all of the original Seven Sisters that dominated world petroleum production after World War II, only now they're organized as four companies. In terms of daily production capacity, the smallest of these companies is six times larger than the original Standard Oil Company that was broken-up in 1911. Collectively they produce 16.8-million B/D, or 19% of the world's oil supply, which is close to the demand value required for the United States economy. A significant factor that impacted the steady rise in oil prices that occurred since 1998 when oil traded at approximately $12 ($17) per barrel is the lack of an appropriate level of competition between major oil companies that produce, refine and market petroleum in the United States.

## 2000's: Radical Islam & the "Oil Price Bubble"

The United States continued to lead the world as the major consumer of petroleum as it entered the new millennium. Petroleum consumption reached 19.8-million B/D of which

9.5-million B/D were produced within the United States. Import volumes reached a reckless 10.3-million B/D.

After the events of September 11, 2001, Americans learned the hard way that Shiite radicals were a moderate and practical form of Islam compared to the Wahhabi Sunnis. A strange thing happened after the destruction of the World Trade Center Towers in New York City, the price of oil continued a steady decline throughout the year 2001. West Texas Intermediate, one of the crude oil benchmarks, entered the year valued at $28.54 ($38.29) per barrel. Its September 2001 value was $24.56 ($32.90) and it continued to decline each month until it dropped to $18.51 ($24.80) by December 2001. The events of 9/11 caused a serious contraction in the size of the United States economy with a related recession in the fourth quarter of 2001. Usually bad news signals reactionary trading that leads to dramatic price spikes, but in this case the financial district where the trades would have been made was wrecked. Oil prices would not increase again until the financial trading sector could re-organize later in 2002.

Earlier in the book we learned about the steady increase in the cost of oil that began in 2002 and ended in the "Price Bubble" event of 2008. Paper trading in the futures market is a significant factor that drives up oil values when you consider there are 15 paper trades for every real barrel of oil. Another factor that can't be ignored is the value of the funds that pay for the oil. Oil is traded in the value of the United States Dollar. When the value of the USD declines, the value of oil goes up accordingly, and when the USD value increases, the value of oil declines. The first decade of the 21st Century witnessed a steady decline in the value of the US Dollar and an associated increase in the value of petroleum until the "Price Bubble" burst in mid-2008.

## 2010's: Alternative Energy

As the United States entered the second decade of the 21st Century, petroleum consumption was actually 300,000 B/D lower than the decade before at 19.2-million B/D, with domestic production of 9.8-million B/D; imports were 9.4-million B/D. Significant contributions to the drop in consumption are related to new and emerging sources of energy such as wind and solar power for the generation of electricity as an alternative to fossil fuels; synthetic fuels for automobiles such as ethanol and bio-diesel; and electric or hybrid-electric automobiles

The United States has turned the corner on two issues of major concern related to petroleum. The first is the realization that the United States economy has the ability to expand without increasing the need for more petroleum because of the shift toward alternative energy sources. The second reality is that bad news from the Middle-East has a significantly lower impact on the price of oil than it did in the 1970's.

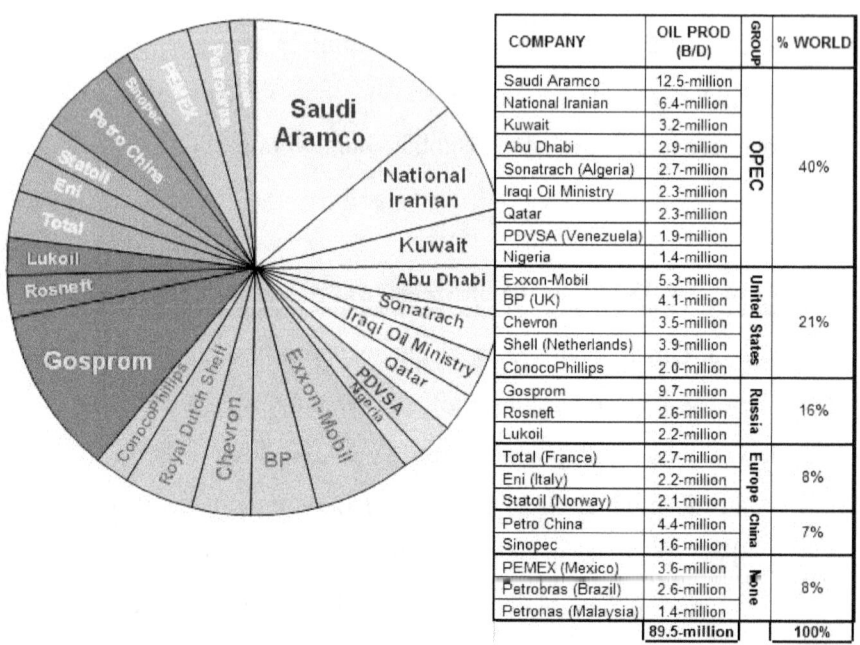

| COMPANY | OIL PROD (B/D) | GROUP | % WORLD |
|---|---|---|---|
| Saudi Aramco | 12.5-million | | |
| National Iranian | 6.4-million | | |
| Kuwait | 3.2-million | | |
| Abu Dhabi | 2.9-million | | |
| Sonatrach (Algeria) | 2.7-million | OPEC | 40% |
| Iraqi Oil Ministry | 2.3-million | | |
| Qatar | 2.3-million | | |
| PDVSA (Venezuela) | 1.9-million | | |
| Nigeria | 1.4-million | | |
| Exxon-Mobil | 5.3-million | | |
| BP (UK) | 4.1-million | | |
| Chevron | 3.5-million | United States | 21% |
| Shell (Netherlands) | 3.9-million | | |
| ConocoPhillips | 2.0-million | | |
| Gosprom | 9.7-million | | |
| Rosneft | 2.6-million | Russia | 16% |
| Lukoil | 2.2-million | | |
| Total (France) | 2.7-million | | |
| Eni (Italy) | 2.2-million | Europe | 8% |
| Statoil (Norway) | 2.1-million | | |
| Petro China | 4.4-million | China | 7% |
| Sinopec | 1.6-million | | |
| PEMEX (Mexico) | 3.6-million | | |
| Petrobras (Brazil) | 2.6-million | None | 8% |
| Petronas (Malaysia) | 1.4-million | | |
| | 89.5-million | | 100% |

**Figure 14.  The 25 largest oil companies in the world**

For example, a new terrorist group named the "Islamic State of Iraq and the Levant", or ISIS, is disrupting oil production and distribution in the northern sector of Iraq; and Iran has sanctions restricting the export of oil related to its nuclear policy. In-spite of these mid-east supply disruptions oil prices continue a gradual descent throughout the second half of year 2014. By the time this book goes to print the United States will be the number one petroleum producer in the world. A reduced need for foreign oil will have a very positive impact on price and sensitivity to foreign events.

Forbes Magazine recently published an article on the Internet titled, "The World's 25 Biggest Oil Companies". Figure 14 is compiled from the content of that article and lists these companies in affiliated groups with reference to their share of the world's oil they control.

*Only 25 companies control the world's oil supply.*

OPEC retains majority control with 40% of the world's daily production, but US affiliated companies are the second largest group with 21%. Royal Dutch Shell and BP are listed as virtual United States companies because of their conspicuous presence as retail vendors and vertical integration throughout the country. Russian companies represent the third group with 18% of the world daily production.

World production of oil for year 2013 is published by the US federal government at 75.2-million B/D. The 25 companies referenced in figure 16 have a total production capacity of 89.5-million B/D. The math is simple, if these numbers are correct there's an apparent world petroleum oversupply of 14.3-million B/D.

There are three conclusions that can be made from understanding the significance of this information. One, there's more oil supply than demand throughout the world. Two, only 25 companies control the world's oil supply. Three, with the exception of 1973 Oil Embargo, the formation of the OPEC cartel is good for world petroleum consumers. There needs to be more competition among major oil operators throughout the world.

# 10

*Supply-Demand Dynamics*

---

The most effective way to explain supply-demand dynamics is to demonstrate the process with a simulated model that represents the real world. Consider a town of approximately 40,000 people. You might think of the town you live in; or a small section of a larger community. This hypothetical town or community has approximately 10,000 automobiles and four fueling stations.

For the purposes of this model we shall consider the average automobile to have a fuel capacity of 15 gallons with knowledge that smaller cars generally hold 12 gallons, and larger cars and trucks exceed 20 gallons. The total capacity for the four fuel stations in this model shall be 120,000 gallons which are serving 10,000 automobiles that

103

have a total capacity of 150,000 gallons, if all the fuel tanks were full.

| FUEL STATIONS (1000 Gals) | | AUTO FUEL TANKS (1000 Gallons) | | | | | | | | | | | | | | | |
|---|---|---|---|---|---|---|---|---|---|---|---|---|---|---|---|---|---|
| | | FULL | | | 3/4ths | | | HALF | | | 1/4th | | | EMPTY | | | |
| | | 150 | 140 | 130 | 120 | 110 | 100 | 90 | 80 | 70 | 60 | 50 | 40 | 30 | 20 | 10 | |
| FULL | 120 | (30) | (20) | (10) | - | 10 | 20 | 30 | 40 | 50 | 60 | 70 | 80 | 90 | 100 | 110 | |
| | 110 | (40) | (30) | (20) | (10) | - | 10 | 20 | 30 | 40 | 50 | 60 | 70 | 80 | 90 | 100 | |
| | 100 | (50) | (40) | (30) | (20) | (10) | - | 10 | 20 | 30 | 40 | 50 | 60 | 70 | 80 | 90 | |
| | 90 | (60) | (50) | (40) | (30) | (20) | (10) | - | 10 | 20 | 30 | 40 | 50 | 60 | 70 | 80 | |
| | 80 | (70) | (60) | (50) | (40) | (30) | (20) | (10) | - | 10 | 20 | 30 | 40 | 50 | 60 | 70 | |
| HALF | 70 | (80) | (70) | (60) | (50) | (40) | (30) | (20) | (10) | - | 10 | 20 | 30 | 40 | 50 | 60 | |
| | 60 | (90) | (80) | (70) | (60) | (50) | (40) | (30) | (20) | (10) | - | 10 | 20 | 30 | 40 | 50 | |
| | 50 | (100) | (90) | (80) | (70) | (60) | (50) | (40) | (30) | (20) | (10) | - | 10 | 20 | 30 | 40 | |
| | 40 | (110) | (100) | (90) | (80) | (70) | (60) | (50) | (40) | (30) | (20) | (10) | - | 10 | 20 | 30 | |
| | 30 | (120) | (110) | (100) | (90) | (80) | (70) | (60) | (50) | (40) | (30) | (20) | (10) | - | 10 | 20 | |
| | 20 | (130) | (120) | (110) | (100) | (90) | (80) | (70) | (60) | (50) | (40) | (30) | (20) | (10) | - | 10 | |
| EMPTY | 10 | (140) | (130) | (120) | (110) | (100) | (90) | (80) | (70) | (60) | (50) | (40) | (30) | (20) | (10) | - | |

**Figure 15. Hypothetical automobile fuel tank v fuel station capacity**

Figure 15 is a matrix that illustrates what happens as collective human behavior response impacts supply and demand dynamics. The positive values (upper right) in the matrix represent the volume of surplus fuel that exists when the vendors have more fuel capacity in their storage tanks than the collective volume of all the automobiles in the community. The negative values (lower left) represent the deficit that would occur if the automobiles have more fuel in the sum of their tanks than the vendors.

The first point that needs to be understood from analysis of the matrix is the fact that there is far more fuel capacity in the 10,000 automobiles than the four fuel stations. Now consider the community you live in . . . that's right, the collective automobile fuel capacity of your local community exceeds the capacity of the fuel stations. So how come they don't run out of gas? Those of us that are old enough to remember realize that in 1973 and 1979 the local fuel stations did run out of gas. It may have happened a few other times since.

There's an adequate supply of fuel in this hypothetical community, or your community, so long as the average automobile is less than half-full and the local fuel stations are greater than half-full. The system remains in balance so long as the collective response of the typical driver is to resist filling his tank before he really needs to, in contrast to the collective response of the venders which is to keep their fuel storage tanks as full as possible.

One of the controls a fuel vender has to reduce demand is changing the price. If he thinks the fuel supply truck will be late he might increase the price to drive away customers until it arrives. If fuel sales are slow he can reduce the price, if operating margin allows. Inner city and airport fuel stations often have fuel prices well above the norm for a region. These are high traffic areas where they have no choice but to put the fuel price beyond the psychological reach of the typical driver or they would rapidly run out of fuel.

Behavior of the collective conscience can create supply disruptions. Refer back to figure 15 to where the number

40(000) in the horizontal, x-axis, meets with number 90(000) in the vertical, y-axis. This point in the matrix represents a vendor surplus of 50,000 gallons. Now let's consider the impact of a collective driver response.

Suppose a news event filters through the community that suggests a supply disruption will occur in the very near future. The natural response to such news is that drivers will shift from a strategy of resistance to filling their fuel tanks to one of keeping them full. As drivers fill their tanks the matrix condition shifts to the left until the fuel stations hit condition red; they've run out of fuel. There is no possibility for the four fuel stations to maintain their supply if all or most of the drivers in the community chooses to keep their automobile tanks full.

Behavior of the collective conscience can create supply disruptions. Supply disruptions result in price increases or rationing. People like us have a share in the responsibility of what happens based on how we choose to respond to the news.

# 11

## *Petroleum Refining*

Each refinery is designed to take a specific type of oil for input in order to create a variety of consumer product outputs. Refinery processes are complicated, but in its simplest terms the crude oil is heated to 370°C (700°F) and pumped into a vertical tank called a distillation tower. Natural gas is bled from the top of the tower, followed by gasoline, naptha, kerosene, diesel, lubricating oil, and heavy oil, in that order from a series of valves descending down from the top; asphalt is extracted from the bottom of the tower, illustrated in figure 16.

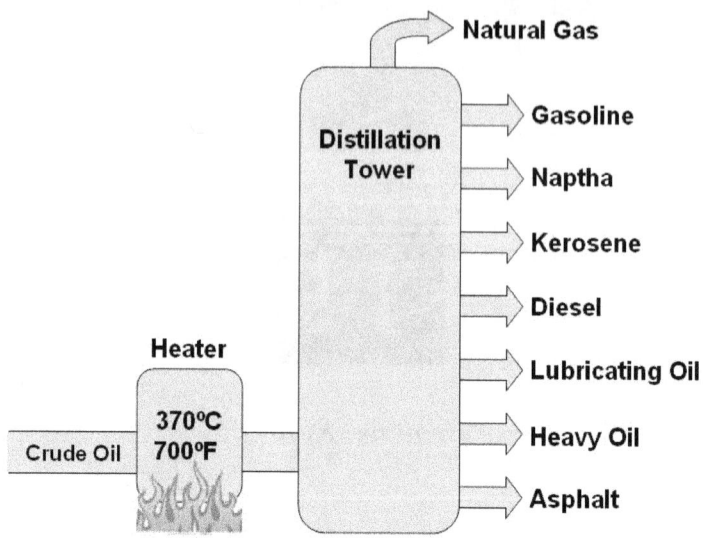

Figure 16.  Simplified schematic of refinery distillation tower

A summary composite of the products produced from a barrel of crude oil compiled from data published by the United States federal government for year 2014 is depicted in figure 17. A 42 gallon barrel of oil will typically produce 44 gallons of refined products; that's right, there's more output than input. The volume of high priority products such as gasoline is enhanced by the catalytic cracking of larger petroleum molecules. Feedstock molecules can be integrated into larger molecules of higher complexity such as plastics and composites by the polymerization of smaller, simpler petroleum molecules.

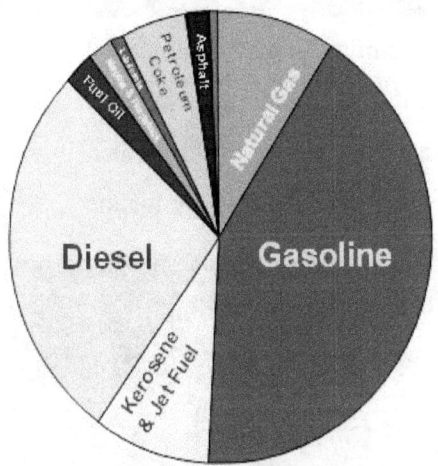

| REFINED PRODUCT | VOL |
|---|---|
| Natural Gas | 4.0 |
| Gasoline | 18.4 |
| Kerosene & Jet Fuel | 3.86 |
| Diesel | 12.1 |
| Fuel Oil | 1.15 |
| Naptha & Feedstock | 0.82 |
| Lubricants | 0.41 |
| Petroleum Coke | 2.18 |
| Asphalt | 0.86 |
| Miscellaneous Products | 0.25 |
| **TOTAL (Gallons)** | **44.0** |

**Figure 17. Summary composite for refined products from 1 barrel crude oil**

There are 142 operable refineries in the United States in year 2014, down from the 300 that existed in 1980. These refineries are spread throughout 41 of the states. Only the states of Florida, Maryland, New York, North Carolina, Virginia, Missouri, Nebraska, Arizona, and Oregon do not have petroleum refineries. The combined production capacity of these refineries is 18.9-million B/D, which is slightly below the national demand value of 19.2-million B/D.

## Benchmark Crude

Each region of the world has premium grades of crude oil used for determining the trade value in commodities markets termed benchmarks. West Texas Intermediate (WTI), which is graded as light and sweet, is the benchmark crude for the United States. Light crude has a low density and is optimal for producing gasoline. Although the margin is relatively small, the production of gasoline is very profitable

> *People like us have a share in the responsibility of what happens based on how we choose to respond to the news.*

because of the volume sold to meet the demands of the consumer market. Sweet crude has a sulfur content below 0.5%. WTI contains 0.24% sulfur, very low by world standards. In contrast to sweet crudes, sour crudes have reduced value because of the potential for damage they impart to refinery equipment.

Brent crude is the primary international benchmark, generally destined for European refineries; graded light, sweet crude. It is not as light as WTI and has a higher sulfur content of 0.37%. Although somewhat inferior, Brent crude has maintained a higher commodity value than WTI since 2010. The two benchmarks diverged after a glut occurred when a series of US refineries went offline and WTI experienced a period of unsold surplus. The price divergence is amplified by the fact that United States oil is not allowed to be exported because of restrictions placed on the petroleum industry as a result of the Energy Policy and Conservation Act (EPCA) of 1975.

Brent crude is a portable commodity that can be exported to anywhere in the world. WTI crude is a restricted commodity that can only be sold in the United States. The contrast in portability contributes to the value contrast. The EPCA is an obsolete law that needs to be repealed. It may have been good policy, or more likely irrelevant, in the 1970's when the United States was facing petroleum shortages. Repeal of the law would not increase the value of WTI, but more likely drop the value of Brent to parity with WTI; and that reality would be good for petroleum consumers.

## Gasoline

The United States appetite for gasoline is 9.4-million B/D for the year 2014, and represents 50% of the refined petroleum products used in the United States. Figure 18 is a chart that removes the mystery behind gasoline prices starting with the price of crude oil in the range from $35 to $150 per barrel. Other factors represented in the chart that impact the price of gasoline includes refining costs (10%), distribution and marketing (5%), operating margin (3%), otherwise known as profit, and of course taxes. Federal tax is a fixed $.184 per gallon. State taxes vary from $.308 in

Alaska, to $.685 in the state of New York.

The table is created with the national average state tax of $.496 per gallon. Federal and state tax collections are approximately ten times the value of refinery profits. The price of gasoline in remote regions or high traffic areas does not follow the graphic. States such as Alaska or Hawaii pay more for gasoline because of additional refining and distribution costs. California has higher gasoline prices due to the cost impact of environmental politics.

| BARREL PRICE | $115.00 | $120.00 | $125.00 | $130.00 | $135.00 | $140.00 | $145.00 | $150.00 |
|---|---|---|---|---|---|---|---|---|
| Gallon Price | $ 2.614 | $ 2.727 | $ 2.841 | $ 2.955 | $ 3.068 | $ 3.182 | $ 3.295 | $ 3.409 |
| Refining | $ 0.261 | $ 0.273 | $ 0.284 | $ 0.295 | $ 0.307 | $ 0.318 | $ 0.330 | $ 0.341 |
| Distro & Market | $ 0.131 | $ 0.136 | $ 0.142 | $ 0.148 | $ 0.153 | $ 0.159 | $ 0.165 | $ 0.170 |
| Profit | $ 0.078 | $ 0.082 | $ 0.085 | $ 0.089 | $ 0.092 | $ 0.095 | $ 0.099 | $ 0.102 |
| SUBTOTAL | $ 3.084 | $ 3.218 | $ 3.352 | $ 3.486 | $ 3.620 | $ 3.755 | $ 3.889 | $ 4.023 |
| Fed Tax | $ 0.184 | $ 0.184 | $ 0.184 | $ 0.184 | $ 0.184 | $ 0.184 | $ 0.184 | $ 0.184 |
| State Tax | $ 0.496 | $ 0.496 | $ 0.496 | $ 0.496 | $ 0.496 | $ 0.496 | $ 0.496 | $ 0.496 |
| TOTAL | $ 3.76 | $ 3.90 | $ 4.03 | $ 4.17 | $ 4.30 | $ 4.43 | $ 4.57 | $ 4.70 |
| BARREL PRICE | $ 75.00 | $ 80.00 | $ 85.00 | $ 90.00 | $ 95.00 | $100.00 | $105.00 | $110.00 |
| Gallon Price | $ 1.705 | $ 1.818 | $ 1.932 | $ 2.045 | $ 2.159 | $ 2.273 | $ 2.386 | $ 2.500 |
| Refining | $ 0.170 | $ 0.182 | $ 0.193 | $ 0.205 | $ 0.216 | $ 0.227 | $ 0.239 | $ 0.250 |
| Distro & Market | $ 0.085 | $ 0.091 | $ 0.097 | $ 0.102 | $ 0.108 | $ 0.114 | $ 0.119 | $ 0.125 |
| Profit | $ 0.051 | $ 0.055 | $ 0.058 | $ 0.061 | $ 0.065 | $ 0.068 | $ 0.072 | $ 0.075 |
| SUBTOTAL | $ 2.011 | $ 2.145 | $ 2.280 | $ 2.414 | $ 2.548 | $ 2.682 | $ 2.816 | $ 2.950 |
| Fed Tax | $ 0.184 | $ 0.184 | $ 0.184 | $ 0.184 | $ 0.184 | $ 0.184 | $ 0.184 | $ 0.184 |
| State Tax | $ 0.496 | $ 0.496 | $ 0.496 | $ 0.496 | $ 0.496 | $ 0.496 | $ 0.496 | $ 0.496 |
| TOTAL | $ 2.69 | $ 2.83 | $ 2.96 | $ 3.09 | $ 3.23 | $ 3.36 | $ 3.50 | $ 3.63 |
| BARREL PRICE | $ 35.00 | $ 40.00 | $ 45.00 | $ 50.00 | $ 55.00 | $ 60.00 | $ 65.00 | $ 70.00 |
| Gallon Price | $ 0.795 | $ 0.909 | $ 1.023 | $ 1.136 | $ 1.250 | $ 1.364 | $ 1.477 | $ 1.591 |
| Refining | $ 0.080 | $ 0.091 | $ 0.102 | $ 0.114 | $ 0.125 | $ 0.136 | $ 0.148 | $ 0.159 |
| Distro & Market | $ 0.040 | $ 0.045 | $ 0.051 | $ 0.057 | $ 0.063 | $ 0.068 | $ 0.074 | $ 0.080 |
| Profit | $ 0.024 | $ 0.027 | $ 0.031 | $ 0.034 | $ 0.038 | $ 0.041 | $ 0.044 | $ 0.048 |
| SUBTOTAL | $ 0.939 | $ 1.073 | $ 1.207 | $ 1.341 | $ 1.475 | $ 1.609 | $ 1.743 | $ 1.877 |
| Fed Tax | $ 0.184 | $ 0.184 | $ 0.184 | $ 0.184 | $ 0.184 | $ 0.184 | $ 0.184 | $ 0.184 |
| State Tax | $ 0.496 | $ 0.496 | $ 0.496 | $ 0.496 | $ 0.496 | $ 0.496 | $ 0.496 | $ 0.496 |
| TOTAL | $ 1.62 | $ 1.75 | $ 1.89 | $ 2.02 | $ 2.16 | $ 2.29 | $ 2.42 | $ 2.56 |

**Figure 18. Retail price calculations for regular unleaded gasoline**

The Octane Number is not related to the quality or energy content of the fuel, but the burn rate. High octane gasoline burns slower than lower octane gasoline. The additional cost associated with high octane gasoline is a product of extra refining with organic chemical ingredients of higher complexity that burn slower than the short and simple hydrocarbons that make low octane gasoline. There is no advantage to using a higher octane fuel than an engine is designed to use. The engine compression ratio (C/R) is the primary parameter that determines the octane number requirement for gasoline:

C/R   8.5:1 = 87 Octane Number
9.5:1 = 92
10.5:1 = 95
11.0:1 = 100

The rating system is based on the concept that octane ($C_8H_{18}$) has a burn rate value of 100. Simpler gasoline molecules burn faster, so they have the potential to create pre-ignition, or engine knocking in high compression engines. It may be counterintuitive but the use of high octane fuel in a low compression engine has adverse effects. High octane fuel is designed to burn at a slower rate, so it has the potential to pass through lower compression engines without burning at all. Non-combustion of fuel in the engine will

113

reduce efficiency and has the potential to cause exhaust system fires. In high altitude regions it is appropriate to use a lower octane number fuel as the lower air pressure and reduced oxygen content in the atmosphere will reduce the burn rate of the fuel.

Thermodynamic efficiency of an internal combustion engine is a product of the difference between the temperatures of intake verse exhaust. High compression engines have higher combustion temperatures and the potential for higher efficiency. Ideally high octane fuel should be used to improve the fuel efficiency of automobiles. Too often it's used to increase the performance of oversize engines in undersize cars. Ultimately efficiency is about reducing operating costs so that high efficiency cars generally use low octane fuel.

# 12

## *Environmental Mishaps*

No book on 21st Century petroleum would be complete without a discussion related to environmental issues. Petroleum exploration and development does not have to be harmful to the environment. We learned in the Introduction that petroleum saved the whales from extinction in the 19th Century. Gulf Coast petroleum platforms are good for the shrimp industry. As artificial reefs they serve as deep water interfaces that enhance the creation of plankton blooms that ultimately leads to improved shrimp populations. Offshore platforms have simplified the shrimp harvesting process, just circle the platforms. That is of course until there's a man induced blunder.

## Deepwater Horizon

On April 20, 2010 the Deepwater Horizon, positioned in 5,067 feet of water, drilled to a total depth of 18,360 feet, hit a methane hydrate pocket, exploded and burned. Eleven members of the drilling crew died and 4.9-million barrels of petroleum discharged into the Gulf of Mexico until the disaster was brought under control 87 days later, on July 15, 2010. A huge international effort was mobilized to support the clean-up effort to the extent that minimal effects remain today. Fish, crabs, shellfish and shrimp populations have returned to normal or near normal for the region. The primary lesson learned from the event is that blow-out prevention in deep water, when drilling to extreme depths, needs to include mitigation for striking methane hydrate deposits.

The federal agency responsible for regulation and oversight of the tragedy received the maximum penalty possible. It was required to re-organize under a new name, formerly known as the Mineral Management Service (MMS), it is now called the Bureau of Ocean Energy Management, Regulation and Enforcement (BOEMRE).

It's a natural reaction for concerned citizens to want to punish the oil company responsible for a disaster of this magnitude by seizing their assets, breaking them up, and compelling the new company to re-organize under the banner of one or many of their competitors. But that didn't happen. BP is the third largest company in the United Kingdom with $351-billion in sales and represents about 15% of the British economy. BP is publicly traded and stock dividends are one of the foundation blocks for the United Kingdom pension system. Economic politics overrule justice and making us feel good. Beyond paying for cleaning up their mess, BP is still very much in business and received nothing more than a glorified slap on the wrist. However, other petroleum companies, not BP, operating in the Gulf of Mexico that had no responsibility in the disaster and take appropriate mitigations for operational risk had their drilling permits held up for over a year.

## Exxon Valdez

It was long enough ago that we forgot the date but March 24, 1989 is the day when a drunken tanker captain ran his ship onto a reef, spilling 750,000 barrels of crude oil into Prince William Sound, Alaska. Twenty-five years later the salmon have returned to normal populations, but the crabs and herring have not. Near Arctic waters are far less resilient to the damage caused by crude oil spills because there is significantly less microbiological activity in cold water and short warm seasons. The lesson learned from this event is that tankers should have double hulls, specifically ones that operate in high latitudes where icebergs are probable. There's no need for new rules prohibiting a ship captain from operating drunk and driving into a reef, the one's that already exist didn't work and additional rules would be redundant.

## Ixtoc I

Thirty years before Deepwater Horizon, there was the IXTOC I, owned by PEMEX, the national petroleum company of Mexico. The most tragic element of this event is the fact that it could have been avoided if the drilling platform, operating north of the Yucatan Peninsula, had a more experienced crew. The platform mounted rig lost mud circulation on June 2, 1978 when drilling at 11,800 feet. The traditional solution to lost circulation is to increase the weight of the drilling mud and to keep the drill string in the hole as this is the heaviest component in the system. With the drill string removed the mud wasn't heavy enough to keep the hole closed. On June 3, 1979 drilling mud spewed out of the hole, followed by a giant plume of burning natural gas, followed by 3.3-million barrels of oil over the next 293 days. Much of the oil washed up on Texas beaches. The IXTOC I blow-out was brought under control on March 23, 1980. Studies related to the long term effects of the event are inconclusive and continue to be an issue of debate.

The lesson from IXTOC I is that government owned companies need to have experienced professional work crews just like private companies, even better than BP. But PEMEX claimed sovereign immunity, and paid little in damage settlements for all the mess they made.

## World War II

Over 6,000 ships were lost at sea due to naval and air battle damage throughout the oceans of the world from 1939 through 1945. Although specific records fail to exist, a realistic estimate for the volume of petroleum lost could be 10-million barrels of crude and refined fuel products. World War II is clearly the most violent period in history in reference to ships lost and liquid petroleum cargos dumped into the sea; yet there are no known long term effects recorded or remembered; history is silent.

There are two lessons from the experience of World War II. The oceans are resilient, and ultimately environmental effects are based on cultural perspectives. If a war of that magnitude were to be fought today, battle reporting would focus on three elements, human life, target objects such as ships, and damage to the environment. This last item wasn't part of the record 70 years ago.

## National Environmental Protection Act

In summary, there are three levels of environmental readiness documentation associated with the National Environmental Protection Act (NEPA). The Categorical Exclusion (CATEX) requires the lowest level of documentation. Generally less than 10 pages in length and less than a week of preparation time; the CATEX is used when the intended action is considered to have no possibility for adverse impact on the environment.

*BP is still very much in business and received nothing more than a glorified slap on the wrist.*

The second level of NEPA documentation is called an Environmental Assessment, or EA. The EA is generally under 500 pages in length and will take a team of skilled persons a month or more to complete. If a review determines that the intended action will result in no significant impact then a Finding of No Significant Impact (FONSI) is issued and the project moves forward. If the potential for significant impact is determined then a full blown Environmental Impact Statement (EIS) is required.

The EIS is a huge set of documents, up to 2,000 pages in length, and requires a large team a year or more to complete, plus an additional four years of review by all interested parties to include the general public and environmental protection groups. Any project that is

120

sensitive about time, schedule, and risk factors wants to avoid the creation of an EIS; that means every petroleum operation wants to avoid the EIS.

The NEPA process is not designed to stop projects that have a potential to impact the environment. The process is intended to slow projects down and give stakeholders, concerned citizens, and environmental advocacy groups an opportunity to provide oversight and opportunities to obstruct the schedule. This can be a good thing.

The Deepwater Horizon operated under a CATEX; and that's a travesty. There is no possible way that a platform rig operating in 5,000 feet of water, and drilling to 18,000 feet, with the possibility of striking methane hydrate deposits, had no potential for environmental impact.

# 13

## *Renewable Energy*

Renewable energy is making positive contributions to the United States energy supply, primarily for electric power generation. Every unit of power used to generate electricity by use of renewable energy sources reduces our dependence on petroleum, which in turn reduces our need for petroleum imports. The standard unit of power for electric energy generation is the Megawatt-hour (MWh), but this book will continue to translate all energy units into the barrel of oil equivalent, or BOE, based on the calculation that one barrel of oil is equivalent to 1.75 MWh. In year 2011 renewable energy accounted for 9% of the United States energy supply, or 4.1-million BOE/D. Wind and solar power are the emerging technologies and account for 540,000 BOE/D and 42,000 BOE/D respectively.

By year 2024 renewable energy is projected to be 5.2-million BOE/D with wind power generating 1.6-million BOE/D and solar power at 360,000 BOE/D. But the current energy requirement for the generation of electric power in the United States is 2.2-million BOE/D. If this trend is maintained, there could be a renewable energy surplus within the next ten years.

## Wind Power

The installation cost of wind power systems is now cheaper than fossil fuel plants and once it's in place there are no fuel costs, day or night, rain or shine, the wind is always free. But wind power can never be more than an augmentation to traditional electric generation because it is intrinsically intermittent. To make wind power feasible it needs to be hooked to a higher capacity electric grid system than currently exists throughout the United States. These upgrades are in work in many communities, paid for through additional fees on your electric bill.

Wind power has adverse impact on the environment as it kills birds and bats that are struck by the blade or explode in the low pressure zones created by the blade vortices. The environmental establishment seems to be looking the other way, as they'll obstruct an energy project that endangers snail darters, but encourage one that destroys bald eagles.

How can the public take these people serious?

New wind power designs are in development that use spiral blades that spin around a vertical axis in contrast to the current designs that use rotary blades that spin around a horizontal axis. The vertical spiral designs are supposed to be safer for birds and bats. It's a good thing snail darters don't fly.

## Solar Power

The opening point that needs to be understood is that solar power is even more intermittent than wind power. It only works well on bright sunlight days at or near noon. Clouds, humidity, snow cover, dust, and pollen reduce solar panel performance. The author of this book has solar panels on his house, not because he's trying to be green, but because he's trying to reduce the impact of the high cost of electricity. The purchase of solar panels was a business decision for the long term welfare of the family. There are a few excellent websites for solar power that attempt to be as informative as possible such as:

> http://pveducation.org
>
> solardat.uoregon.edu.

But it should be no surprise to consumers that the solar industry is based on misinformation as much as any other.

The major source of misinformation within the solar industry is related to optimal sun conditions. Many websites describe how summer is optimal for solar energy generation because of high sun angles and long daylight hours. They commonly reference the solar energy potential of 1,000 watts per square meter as a holy grail that is vital to effective photon collection. All this is partial fact, but incomplete. So what are the facts about solar power?

Photovoltaic panels crave cold sunlight on clear winter days. That's right! Solar panels don't behave like oak trees. If it's warm

> *It should be no surprise to consumers that the solar industry is based on misinformation as much as any other.*

enough to encourage tree growth, it's too warm to get optimal efficiency from solar panels. The current state-of-the-art for residential grade solar panels is approximately 16% efficiency. That means that a solar panel of 1.675 square meters, the standard for the industry, will collect 265 watts of power when aligned orthogonal to the sun; but that power output is only possible if the air temperature is 0°C (32°F). As air temperature increases the efficiency of the solar panel decreases approximately 0.5% per 1°C (1.8°F) as the air temperature rises above 0°C (32°F). The effect of summer temperatures will reduce a 265 watt solar to 220 watts of power output under ideal sunlight conditions when the air temperature is 30°C (86°F). When the oak tree is happily soaking up the sunlight, the solar panel is not.

Another fallacy is related to the myth of long daylight hours. A fixed solar panel will not collect sunlight significant for power generation longer than 9.5 hours. So the fact that summer daylight may exceed 14 hours or more hours is irrelevant to the production of solar power electricity. There's a fixed maximum daylight potential regardless of how many daylight hours exist where you live. So what is the lesson?

Early spring and late fall, in regions where there are cold air temperatures and 9.5 hours of bright sunlight, can be optimal for solar power generation, if the solar panels are placed at a high enough angle to match the incidence angle of the sun at solar noon. But too many solar websites lead us to believe southern locations are optimal for solar power.

The solar industry of today is where the automobile industry was 110 years ago. In the year 2014 the solar industry only sells to the wealthy. There's an irony in this reality. Typically the return on investment for a solar system purchase is 8 to 12 years. Anyone wealthy enough to purchase solar power at the current market price is too wealthy to need the cost benefits. Residential solar power is a true luxury item. The cost of a solar power system is generally two or more times the cost of the parts. The industry is waiting for an innovator that drops the cost of installation in order to dramatically increase sales volume, even to the middle class. It's what Henry Ford did for the automobile industry.

The cost of parts is not the issue. Premium grade solar panels cost under $300 each. When you consider the components and complexity within a solar panel it would be difficult to believe the price of parts could drop significantly lower than it already is.

The efficiency of solar panels has increased an average of 0.45% per year since 1976. If this efficiency trend is maintained, solar panels

*Anyone wealthy enough to purchase solar power at the current market price is too wealthy to need the cost benefits.*

have the potential to be 28% efficient by year 2040. Solar power will need this efficiency increase as the industry is heavily subsidized by the federal government. Some states offer solar energy credits or even solar renewable energy credits that pay residential home owners for the energy they produce in addition to a reduced electric bill. The industry needs these subsidies to survive today but these subsidies won't last forever. Nor should they, the industry will need to increase efficiency and reduce installation costs if it is to survive long into the future.

## If Cows Produced Gasoline

Similar to the chicken that does a wonderful job converting insects and corn feed into eggs, the common dairy cow is very efficient at converting grass, clover, and fodder into milk. The dairy cow is able to convert 27 pounds of forage into 6.6 gallons of milk each day.

Let's use our imaginations and consider a world where an animal similar to the dairy cow was able to produce a liquid hydrocarbon similar to gasoline. With the current United States consumption rate for gasoline at 9.4-million B/D, it would take 60,000,000 gasoline cows, let's call them gows, to supply this nation with the renewable fuel needed to replace petroleum based gasoline. Using prime pasture east of the Mississippi this number of gows would require 187,000 square miles of land, or three times the area of Wisconsin.

Now let's consider ethanol from corn. The national average for the corn harvest is 159 bushels per acre. Each acre can produce 61 gallons of ethanol, or 1.5 BOE. The United States requires 9.4-million B/D, or 3.4-billion barrels per year if ethanol is to replace gasoline. We will not take into consideration that ethanol would require at least 5% more fuel than gasoline because it burns less efficiently. The amount of land needed to replace the United States gasoline market with renewable corn ethanol would require 2.3-billion acres of land, or 3.7-million square miles. That's 25%

more land area than the contiguous lower 48 states.

One of the dangers of using agricultural crops as a fuel source is the competition for the farm land that is traditionally used for food. As cropland for food is converted to cropland for fuel, the cost of both types of commodities will increase until mankind can afford neither. The price of corn and wheat has tripled in price since 2005 when ethanol fuel started to become common as a fuel additive.

Is this really what we want? The best source for hydrocarbon fuel is the one we're already using, petroleum. It's counterintuitive but energy sources such as petroleum, which appear finite, have an inexhaustible volume of supply, and renewable energy sources, which appear infinite, suffer from a lack of supply.

# 14

## *Facing Challenges*

---

The key to understanding energy and related environmental policies is to understand the motivations behind the players engaged in the issues. That can be a challenge. Technical challenges can be solved through logic and analysis that leads to science and engineering based solutions. Energy and the environment seems like it should be a technical challenge. But in truth, understanding human interactions and motivations are a mystery to those trained in technology; unless they understand the Solution Approach Matrix, figure 19.

There are two axes: Commitment to Solving the Problem and Creativity Level. Commitment to solving the problem can be weak or strong; creativity level is traditional or visionary. Traditional creativity with weak commitment leads to reactive solutions where the players are motivated to maintain the status quo. Traditional creativity with strong commitment leads to brute force solutions. Visionary creativity with weak commitment leads to gimmicks and visionary creativity with strong commitment leads to proactive solutions.

Programs or policies that remain in a constant state of controversy are operating in the REACTIVE box where maintenance of the status quo is the priority. Now it's a natural question to ask, how can this be . . . why would a program or policy want to remain in a constant state of controversy? The solution to all such questions is to follow the money trail. Lawyers and bureaucrats own and control the REACTIVE box. The NEPA regulation process that is perceived to exist to resolve environmental issues, in reality is used to guarantee issues of concern never quite get resolved. The process was developed for and by lawyers and leads toward a huge income stream for litigation agents on both sides of any issue.

| SOLUTION APPROACH MATRIX | | |
|---|---|---|
| | COMMITTMENT TO SOLVING PROBLEM | |
| | WEAK | STRONG |
| **CREATIVITY LEVEL** — **VISIONARY** | GIMMICKS - Superficial feel good solutions that seem positive but don't really solve the problem | PROACTIVE - New S&T developments to improve state-of-the-art. Results from original thinking and seeks creative solutions |
| **TRADITIONAL** | REACTIVE - Protect the *status quo*, likely to maintain "high concern/low trust" environment | BRUTE FORCE - spend more other people's money based on a transfer of priorities |

**Figure 19. Solution Approach Matrix**

To understand how and why issues get stuck in the status quo we need to review a wonderful book by Frances Eliza Hodgson Burnett published in 1910 titled, *"The Secret Garden"*. This is the story of a ten year old boy that is crippled and sickly and is led to believe that he will soon die and never learn to walk. There is a lady character that serves as a housekeeper nurse whose livelihood it is to care for the boy. She believes her charter is to keep the boy sick and crippled in spite of the fact she is employed to improve the boy's health. The boy has a father that has troubles of his own and so he is preoccupied to the extent that he doesn't provide appropriate supervision to the nursing of his son or gain a real understanding of his true condition or future potential.

The sick boy has a cousin that enters his life and everything changes. Together they discover a secret garden in disrepair and restore it to its former beauty. The girl is hated by the housekeeper nurse and is ignored by the sick boy's father. In summary, the housekeeper nurse does everything in her power to keep the two cousins apart. When they meet the boy quickly regains his health, learns to walk, and even learns to run. His father returns from his travels and they all live happily ever after. The most fascinating component of the story is the fact that the housekeeper nurse feels powerless and unworthy when the boy gets his health and strength and acts like a real boy in spite of the fact she was hired to care for the boy and bring him to health.

This story my dear readers is a perfect metaphor for the plight of our national energy strategy and its impact on our economy! The housekeeper nurse is the government bureaucrats in league with the major oil companies that control energy policies in the United States. The sick boy is the American citizens that suffer because of the lies, politics, and misinformation that drive our dysfunctional economy and steals wealth and opportunity from energy consumers. The secret garden is the United States economy that suffers from high energy costs. The boy's father and Lord of the Manor are the political leaders that are empowered to solve national issues of concern but are preoccupied with other business. The girl represents the creative science and technology developers that could quickly return our nation to a vital economy if only they were given the opportunity to do so.

What is the message? If we are committed to a functional national energy policy we need to get rid of the nurse or make her irrelevant. We need to empower the girl and get the nurse out of her way. Just like the boy in the story that learned to walk and run when given his opportunity, our economy can recover to become the world's major energy supplier again.

So what's the solution? BRUTE FORCE solutions result from adding to the level of commitment but maintaining a traditional level of creativity. Government agencies and large corporations with lots of other people's money to spend often implement this approach. The brute force process is generally a waste of resource as people, materials, or funds are taken from one sector and given to another under the justification that one is more important than the other. Mass transit systems are often a brute force solution that cost too much and don't provide appropriate benefits to the consumers that are compelled to pay for them against their will.

Shale gas extraction by use of fracking has elements of brute force. Although the technique improves gas production in source beds that would otherwise be dormant, it uses horrendous volumes of water that can't be safely disposed of without damaging watersheds or causing seismic events. There has to be a better way.

GIMMICKS are a product of visionary creativity combined with a low commitment to real solutions. People skilled in the dark side of human psychology, otherwise known as con men, operate in this box with a goal to helping us feel good, without really solving the problem. Cap-and-Trade for carbon credits is very much a gimmick. A loud and well connected minority of insiders feel really well about the prospect of designating trees in one forest as sequestration agents so that remote industrial processes can spew out green house gases. Not all green house gases are considered, just carbon dioxide. The whole system is based on lies and hypocrisy. Carbon dioxide isn't a pollutant and shouldn't be regulated. In the Cap-and-trade system owners of the designated trees would be compensated in spite of the fact there are billions of other tree owners in the world who are not. Who's speaking for the trees? They're crying out for more carbon dioxide and want humans and their machines as oxygen sequestration agents. But nobody is listening.

Real solutions are created when there is a true commitment to solving problems combined with a visionary level of creativity. The result is PROACTIVE solutions that are generally a product of new technology or a process that reduces costs, and creates win-win relationships for all groups engaged in the issue; except for the people that

135

benefited from the status-quo, or brute force, or gimmicks. Proactive solutions are the most difficult to achieve because they require some combination of risk capital, intellectual property development, and turf battles with the groups that would lose wealth and power if the new technology succeeds. Sometimes a new idea is so beneficial to so many groups that the battle never materializes. Petroleum displaced whale oil in the 19th Century. Except for a few whalers that lost their livelihoods, the world never looked back.

Cellulosic ethanol is a technology under development that will convert all forms of vegetation biomass into liquid fuels. If or when the technology is viable there is a potential for a farmer to grow a crop, sell the fruit to the food processors, vendors or consumers, and deliver the leaves, stalks, and other organic waste components to the fuel plant. The capability for cellulosic biomass conversion into fuel from waste is a win-win for farmers and fuel consumers throughout the world. Unlike traditional brute force ethanol conversion that steals food from one process to make fuel for another, cellulosic ethanol is a proactive solution.

## Energy Serfs

Why the focus on carbon dioxide and not real pollutants like sulfur dioxide, nitrous oxides, or mercury compounds?

These real pollutants are ignored because there are environmental laws to bring them under control, are found in trace amounts, and are treatable when existing laws are enforced. Real pollution is not part of the fuel combustion equation so if they're used to scare people the response would be, "fix the problem you have the laws you need".

Carbon dioxide is a product of nature, it can't be avoided, and it's easy to measure. It's too easy to be used as a threat to lead uninformed citizens toward energy serfdom to be ignored by the environmental establishment.

*Proactive solutions are the most difficult to achieve because they require . . . turf battles with the groups that would lose wealth and power if the new technology succeeds.*

The manifest goal of the environmental establishment is to convince us we are destroying the planet by living too well with the latent intent to get us to trade our real energy freedom for false environmental security. The first step in the process toward Energy Serfdom is to convince us to feel guilty about our energy choices such as our car, truck, or carbon footprint is too big; our standard of living is too high; and others live in poverty because we live too well.

If or when we accept smaller cars, smaller homes, and a lower standard of living as the norm, the economy will contract. Meanwhile the group that led us toward these conditions will continue to fly around the world in their personal jets and ride in limousines, or oversize SUVs. Another way to put it, anybody that wants to stand behind a podium and complain about the American way of life better ride a horse to the conference and wear clothing made of nothing but natural fibers such as cotton, wool, linen, or silk, otherwise they're a hypocrite.

# 15

## *Conclusion*

---

As consumers we've been told the world is running out of petroleum and in the first half of the year 2014 the uninformed may have thought that was possibly true. Since June 2014 the price of oil has dropped from over $115 to under $50 per barrel. The oil producers are now willing to admit there's an over supply. This oversupply condition didn't just start within the past year. It's always been with us. As petroleum consumers we've been lied to by the producers in order to keep their profits as high as possible. The mystery is resolved, there's plenty of petroleum and for most of the last century we've been paying too much for it.

Except for the early 20th Century, when President, Congress, and the Judiciary worked as a team to dismantle Standard Oil, this nation has rarely had a national energy policy driven by elected or appointed political leaders. Too often the energy agenda in the United States is driven by loud minorities who have wealth and access, such as oil company executives, or passion, such as the environmentalists.

Real leadership would drive the United States toward 100% domestic energy production, to include Canadian sources. This goal can be achieved by a combination of traditional oil and natural gas; shale oil and natural gas from fracking and horizontal drilling; oil shale extraction

> *The manifest goal of the environmental establishment is to convince us we are destroying the planet . . .*

by the "In-Situ Process"; Canadian tar sand petroleum; and the conversion of coal to synthetic liquid petroleum. In the longer term the United States should work toward extracting natural gas from methane hydrate deposits in the Arctic and deep off-shore. Fischer-Tropsch conversion of natural gas, coal, and methane hydrate to liquid fuels should be a national priority. Dry fracking processes need to be developed.

In no way should renewable energy sources that are economically viable be discouraged. Wind power is over the development cost-risk hump and should be considered in any area where there is adequate wind potential and local community support. Solar power is still at risk, the subsidies

140

can't and won't last forever; and photovoltaic efficiency needs to improve for it to become viable in the long term. The development of cellulosic ethanol needs to be encouraged so we can stop using food crops as fuel.

OPEC is not the enemy to the United States energy independence any more than the major oil companies. Both groups have a right and a responsibility to maximize honest profits without the use of manipulations based on falsehoods. If the profits are too high or the distribution of the resource is unbalanced then the public has a right and a responsibility to bring them under control. The first step toward controlling these powerful groups is to understand that they have plenty of supply and many other energy alternatives long into the future. Petroleum suppliers need to be compelled to compete in order to reduce their prices or get them out of the way. The first step in price accountability is facing the fact that there are plenty of petroleum sources and synthetic equivalents to serve mankind long into the future. We will not run out of petroleum fuel anytime soon.

The United States needs to establish energy policies that make whatever happens in the Middle-East or the rest of the world irrelevant to the fuel supply in North America. This goal is totally achievable. If the major oil companies won't work toward national energy independence, smaller independent oil companies can fill the gap. If the big oil companies continue to resist with anti-trade policies then they can be threatened with a replay of the 1911 break-up of Standard Oil.

Imagine a world where a tanker can sink in the Persian Gulf, or a mid-east pipeline or refinery can blow-up and nothing dramatic happens to the price of petroleum. This reality is entirely possible so long as the United States makes a commitment to petroleum independence. The United States is now the major petroleum producer in the world; but the goal of energy independence will not be complete until petroleum imports are terminated.

As this book goes to print the world price for oil is dropping to discourage United States petroleum independence based on the success of North American production increases from United States shale gas and Canadian tar sands. The potential for coal conversion to synthetic liquid petroleum and enormous methane hydrate deposits that exist throughout the world provide the opportunity for affordable fuel for many generations into the future.

Let's return to answer the questions that were posed at the beginning of the book. Petroleum is not obsolete as an energy source and will continue to meet the energy demands of a growing world industrial base long into the future. The United States has the potential to produce enough petroleum or synthetic alternatives to be independent of foreign energy sources. Recent events have even proven that petroleum can be very affordable. Accept the fact that this planet is blessed with immense petroleum resources and we generally pay too much for it, because it's the truth.

# Information Sources

## Chapter 1 - Introduction

1. http://fineartamerica.com/featured/the-original-1859-
   drake-oil-well-everett.html

## Chapter 2 - Petroleum Basics

1. http://www.firmgreen.com/m/fuel/fuel_facts.htm
2. http://energy.usgs.gov/GeochemistryGeophysics/
   GeochemistryResearch/
   OrganicOriginsofPetroleum.aspx
3. http://www.sepmstrata.org/page.aspx?pageid=299
4. D.R. Statter, "Geology 1010 Lecture Notes", Charles
   County Community College, 1998, p 50-54.
5. "Laboratory Manual in Physical Geology", 4th Edition.
6. http://www.eia.gov/pub/oil_gas/natural_gas/analysis_
   publications/maps/maps.htm
7. http://www.ogj.com/articles/print/volume-111/issue 12/
   special-report-worldwide-report/worldwide-
   reserves-oil-production-post-modest-rise.html
8. http://en.wikipedia.org/wiki/Ghawar
9. http://www.theenergylibrary.com/node/13221
10. http://i129.photobucket.com/albums/p237/1ace11/
    Ghawar1.gif

11. http://gailtheactuary.files.wordpress.com/2013/08/world-oil-price-in-2011-dollars.png

12. http://www.eia.gov/pub/oil_gas/petroleum/analysis_publications/chronology/petroleumchronology2000.htm

13. http://www.eia.doe.gov/basics/quickoil.html

14. http://www.offshore-technology.com/projects/na_kika/

15. http://en.wikipedia.org/wiki/File:758Syms2006OCSMapWithPlanni.png

16. http://www.kusi.com/weather/colemanscorner/38574742.html

17. http://www.wtrg.com/prices.htm

18. http://en.wikipedia.org/wiki/File:Oil_price_chronology-june2007.gif

19. http://www.macrotrends.net/

**Chapter 3 - Natural Gas**

1. http://en.wikipedia.org/wiki/List_of_countries_by_natural_gas_production

2. http://www.pipeline101.com/where-are-pipelines-located/natural-gas-pipelines-map

3. http://www.moneymorning.com.au/20120505/russian-exile-how-europe-will-end-the-kremlins-natural-gas-monopoly.html

4. http://www.eia.gov/dnav/ng/hist/rngwhhdm.htm

5. http://www.iie.com/publications/pb/pb09-19.pdf

6. http://www.eia.gov/energy_in_brief/article/about_shale_
    gas.cfm

7. http://www.eia.gov/dnav/ng/hist/n9100us2A.htm

8. http://en.wikipedia.org/wiki/Dominion_Cove_Point_LNG

9. http://www.eia.gov/dnav/ng/ng_move_poe1_dcu_ycpt-
    nno_a.htm

10. http://energy.gov/sites/prod/files/2013/09/f2/
    Order%203331.pdf

11. http://www.ferc.gov/industries/gas/indus-act/lng/exist-
    term/cove-point.asp

12. http://energy.gov/sites/prod/files/2013/04/f0/
    LNG%20Import%20%26%20Export%20Terminal%
    20Maps%2012-18-2012.pdf

13. Giuliano, "Introduction to Oil and Gas Technology", 2nd
    Ed.1981, Fig 5.37, p 95.

14. http://www.propublica.org/special/hydraulic-fracturing-
    national

15. http://www.eia.gov/energyexplained/index.cfm?
    page=natural_gas_where

16. http://www.eia.gov/dnav/ng/ng_prod_sum_a_EPG0_
    FGW_mmcf_a.htm

17. http://www.forbes.com/sites/peterdetwiler/2014/03/28/
    small-gas-to-liquids-plants-get-a-huge-boost/

18. http://en.wikipedia.org/wiki/File:Gas_Production_from_
    Bakken_2000-2013.png

## Chapter 4 - Other Petroleum Sources

1. http://www.greencarcongress.com/2005/09/
   shell_to_take_6.html
2. http://www.eia.gov/todayinenergy/detail.cfm?id=11611
3. http://csegrecorder.com/articles/view/depth-conversion-
   and-seismic-lithology-inversion-mcmurray-reservoir
4. http://en.wikipedia.org/wiki/Keystone_Pipeline
5. http://www.oil-price.net/en/articles/US-shale-oil-deposits-
   2-trillion-barrels-crude-oil.php
6. Shale oil and shale gas resources are globally abundant –
   Today in Energy – U.S. Energy Information
   Administration (EIA).htm

## Chapter 5 - Coal

1. Basics1-CoalGasification-Jun07
2. IHS-McCloskey-Coal-Report-jan-2014
3. http://en.wikipedia.org/wiki/Fischer%E2%80%93
   Tropsch_process
4. http://en.wikipedia.org/wiki/Coal
5. http://www.arrakis-group.com/energy/coal-sourcing-
   supply-chain-from-us-to-africa/

## Chapter 6 - The Case for Fischer-Tropsch

1. http://en.wikipedia.org/wiki/Oil_tanker
2. http://en.wikipedia.org/wiki/LNG_carrier

3. http://en.wikipedia.org/wiki/Dominion_Cove_Point_LNG

4. http://www.cbi.com/project-profiles/cove-point-terminal

5. http://energy.gov/downloads/dominion-cove-point-lng-lp-fe-dkt-no-11-128-lng

6. Asian LNG Tanker Builders Vie for Market Share – WSJ.htm

7. Order 3331.pdf


**Chapter 8 - Methane Hydrate**

1. http://geology.com/articles/methane-hydrates/

2. http://geology.com/articles/methane-hydrates/

3. http://arctic-news.blogspot.com/2012_05_01_archive.htm

4. https://www.jamstec.go.jp/jamstece/30th/part6/page5.html

5. http://www.sankey-diagrams.com/tag/oil/


**Chapter 9 - United States Petroleum History**

1. http://en.wikipedia.org/wiki/Standard_Oil

2. http://www.opec.org/opec_web/en/

3. Giuliano, "Introduction to Oil and Gas Technology", 2nd Ed.1981, pp4-6.

4. http://www.wtrg.com/prices.htm

5. 2009a_bpea_hamilton

6. Strategic raw materials and oil production, Second World War.htm

7. http://www.forbes.com/pictures/mef45glfe/not-just-the-usual-suspects-2/
8. http://www.virginia.edu/igpr/APAG/apagoilhistory.html

## Chapter 11 - Petroleum Refining

1. http://www.eia.gov/dnav/pet/pet_pnp_cap1_dcu_nus_a.htm
2. http://oil-price.net/dashboard.php?lang=en
3. http://en.wikipedia.org/wiki/Energy_Policy_and_Conservation_Act
4. http://www.eia.gov/petroleum/gasdiesel/

## Chapter 12 - Environmental Mishaps

1. http://www.theguardian.com/environment/2010/may/20/deepwater-methane-hydrates-bp-gulf
2. http://en.wikipedia.org/wiki/Deepwater_Horizon_oil_spill
3. http://en.wikipedia.org/wiki/Exxon_Valdez
4. http://mashable.com/2014/03/24/exxon-valdez-25-years-later/
5. http://www.bbc.co.uk/news/10307105
6. http://en.wikipedia.org/wiki/Ixtoc_I_oil_spill
7. http://www.boemre.gov/

## Chapter 13 - Renewable Energy

1. http://www.nrel.gov/

2. http://www.irena.org/News/Description.aspx?NType
   =A&mnu=cat&PriMenuID=16&CatID=
   84&Ne ws_ID=386

3. http://www.nrdc.org/energy/renewables/wind.asp

4. http://pveducation.org

5. solardat.uoregon.edu.

6. http://www.psisolar.net/

## Chapter 14 - Facing Challenges

1. Frances Eliza Hodgson Burnett, *"The Secret Garden"*,
   1910.

## Chapter 15 - Conclusion

1. http://www.oil-price.net/en/articles/oil-boom-but-
   infrastructure-woes.php

2. http://oil-price.net/en/articles/boom-time-for-crude-oil-in-
   the-us.php

## About the Author

The motivation behind this book was to uncover the great variety of misrepresentations that drive up the cost of petroleum in spite of the fact that it's painfully obvious to informed observers that the world operates with significant production excess even in the years when the price spikes to new highs. Tired of the lies, politics, and misinformation that drive the petroleum industry, Mr. Statter began this book project a few weeks before the mid-year petroleum price peak in June 2014, and finished the first draft when the trend was well into its decline by September 2014.

Mr. Statter is a Petroleum Geologist that had his career displaced by the first Oil Glut as he graduated from college in 1982. He's had a variety of other vocations to include cartographer, college professor, remote sensor analysis and development, and water company president. He resides in Lusby, Maryland, with his wife of over 30 years. He has five children and four grandchildren.

www.ingramcontent.com/pod-product-compliance
Lightning Source LLC
Chambersburg PA
CBHW051917170526
45168CB00001B/434